Cleaving an Unknown World

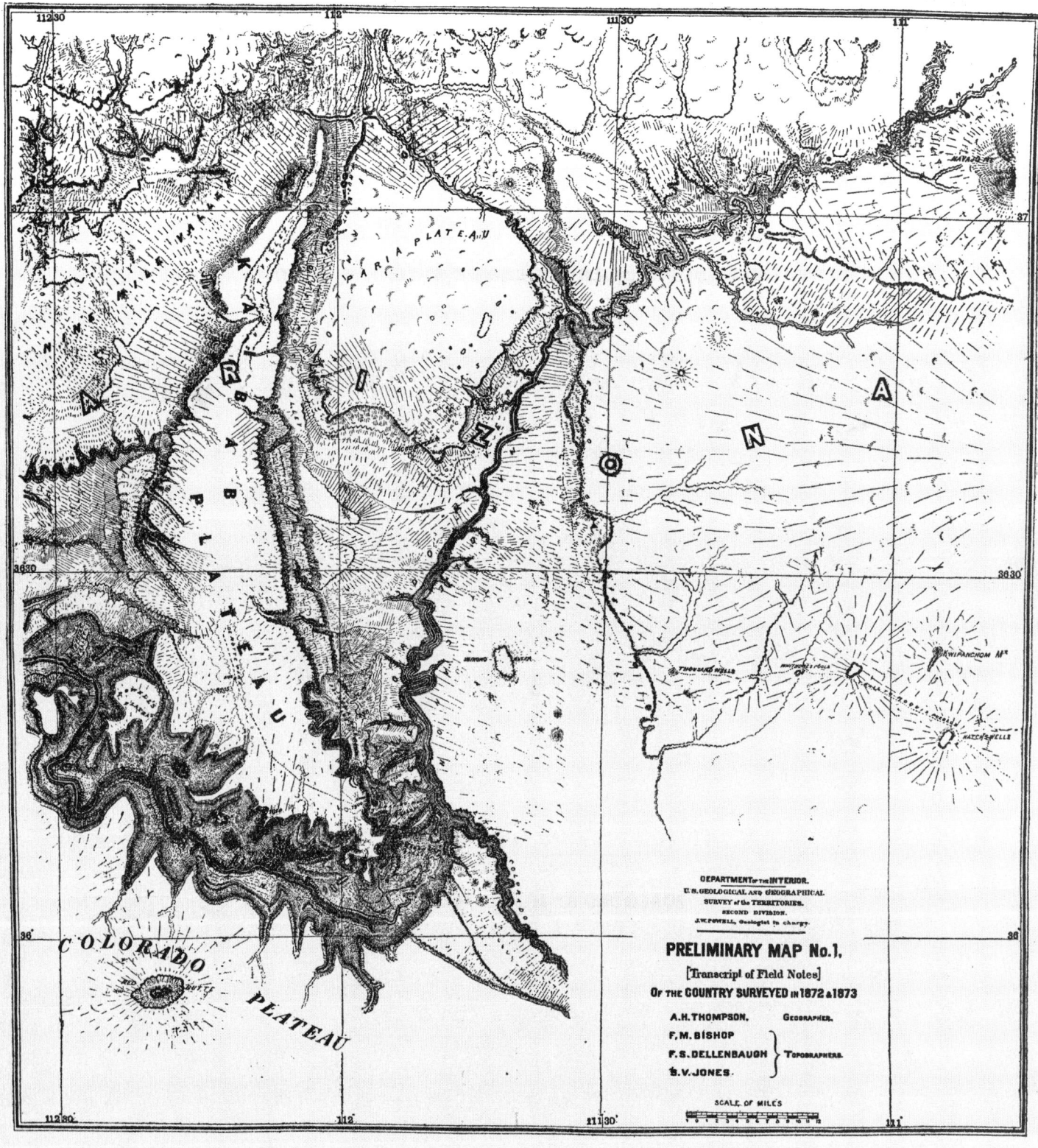

DEPARTMENT of the INTERIOR.
U.S. GEOLOGICAL and GEOGRAPHICAL
SURVEY of the TERRITORIES.
SECOND DIVISION.
J.W.POWELL, Geologist in charge.

PRELIMINARY MAP No.1,

[Transcript of Field Notes]

Of the COUNTRY SURVEYED in 1872 & 1873

A.H.THOMPSON,   Geographer.

F.M.BISHOP
F.S.DELLENBAUGH } Topographers.
S.V.JONES.

SCALE OF MILES

THE POWELL EXPEDITIONS AND THE SCIENTIFIC EXPLORATION OF THE COLORADO PLATEAU

CLEAVING AN UNKNOWN *World*

EDITED BY DON D. FOWLER

FOREWORD BY **ROY WEBB**

THE UNIVERSITY OF UTAH PRESS | UTAH STATE HISTORICAL SOCIETY
Salt Lake City

The Defiance House Man colophon is a registered trademark
of the University of Utah Press. It is based upon a four-foot-tall,
Ancient Puebloan pictograph (late PIII) near Glen Canyon, Utah.

16 15 14 13 12   1 2 3 4 5

LIBRARY OF CONGRESS CATALOGING-IN-PUBLICATION DATA
Cleaving an unknown world : the Powell expeditions and the scientific
exploration of the Colorado Plateau / edited by Don D. Fowler.
p. cm.
Includes index.
ISBN 978-1-60781-146-6 (paper : alk. paper)
1. Colorado Plateau—Discovery and exploration—Sources.
2. Southwest, New—Discovery and exploration—Sources. 3. Colorado Plateau—Description and travel—Sources. 4. Southwest, New—Description and
travel—Sources. 5. Scientific expeditions—Colorado Plateau—History—19th
century—Sources. 6. Powell, John Wesley, 1834-1902—Diaries. 7. Hillers,
John K., 1843-1925—Diaries. 8. Bishop, Francis Marion, 1843-1933.
9. Dellenbaugh, Frederick Samuel, 1853-1935—Correspondence.
10. Sumner, John Colton—Diaries. I. Fowler, Don D., 1936-

F788.C55 2011   979.1'3--dc23
2011026039

Frontispiece: Map C from *A Canyon Voyage*, by Frederick S. Dellenbaugh
Please consider page 243 an extension of this copyright page.
Printed and bound by Sheridan Books, Inc., Ann Arbor, Michigan.

Publisher's Note

Dates of all journal entries have been standardized.
Dates of all letters have been left as original.

CONTENTS

FOREWORD

Roy Webb

In the annals of exploration in North America, the 1869 and 1871 expeditions led by John Wesley Powell down the Green and Colorado Rivers are among the most significant, stirring, and well documented in American history. Powell and his men filled in blank places on the map of the western United States; they met Native Americans who had scarcely seen a white man; they took the first photographs of the canyons; they measured the fall of rapids and the depths of canyons and published them in detailed government reports. And it wasn't just dry description; there was drama, tension, even tragedy and death. There were capsized boats, close encounters with fire, narrow escapes from cliff edges. Powell started with nine crew members in 1869; one departed after a disaster in the Canyon of Lodore, while three others left near the end of the Grand Canyon and tried to climb out, never to be seen again. On Powell's second journey in 1871 and 1872, the travel was just as arduous, the rapids just as dangerous. And all of this was well documented; out of the twenty-one men who were part of the two

crews, no fewer than seventeen kept journals of the expedition or sent long, detailed letters to their families or to hometown newspapers whenever they touched civilization. No other exploring expedition in North American history can be studied from so many and varied viewpoints.

By the time Powell ventured down the Green River in May 1869, most of the continental United States had been explored. Since Lewis and Clark, parties of trappers, U.S. Army surveyors, and emigrants had crisscrossed the continent, from Maine to California, with one exception: the canyons of the Green and Colorado Rivers. Exploration and settlement of the river basins had mostly just nibbled around the edges. Trappers, Argonauts, Mormons, and '49ers had been crossing the upper Green River in what would become Wyoming since the 1820s, while Spanish explorers had come up the Colorado from the Gulf of California as early as the 1500s. But those deep, remote canyons were unexplored and unknown. Writing in August 1843, John Charles Fremont, known as "The Pathfinder" and himself no slouch when it came to mapping uncharted territories, described the state of knowledge about the canyons of the Green and Colorado Rivers.

> The course of the Green and Colorado is but little known, and that little derived from vague report. Three hundred miles of its lower part, as it approaches the Gulf of California, is reported to be smooth and tranquil; but its upper part is manifestly broken into many falls and rapids. From many descriptions of trappers, it is probable that in its foaming course among its lofty precipices, it presents many scenes of wild grandeur; and though offering many temptations and often discussed, no trappers have been found bold enough to undertake a voyage which has so certain a prospect of a fatal termination.[1]

No matter that trappers had ventured into the canyons. William Ashley floated the upper Green in 1825, while James Ohio Pattie had traveled along the north rim of the Grand Canyon around the same time. Joe Meek led a party of trappers that traveled through the Canyon of Lodore on the frozen Green in the 1830s, and the mysterious Denis Julien left inscriptions dated from the 1830s at several places along the Green and Colorado. There were probably others, but their stories were told only around campfires, their accounts and journals unpublished until much later. To virtually everyone living in the interior West, the canyons remained places of great mystery. Rumors abounded that there were impassable waterfalls and cataracts, that the rivers disappeared into caves, that the Lost Tribes of Israel had settled in the canyons. Names like the Green River Suck and the Graveyard of the Colorado were attached to canyons, and everyone knew that anyone fool enough to try to travel down these rivers was doomed from the start.

Into this uncertainty stepped John Wesley Powell. Born in New York in 1834, Powell grew up in rural Illinois. His father, a Wesleyan minister, wanted his son to follow in his footsteps but Wes, as he was often known, was stirred by nature, by the wildlife and plants and geology he saw all around his home, and devoted himself to natural history instead. When the country was rent by civil war, young Powell answered the call and joined an Illinois volunteer regiment, quickly rising through the ranks to become an officer. He served in Ulysses S. Grant's Army of the Tennessee, commanding a battery of artillery. At the Battle of Shiloh in April 1862, Grant was surprised and almost routed by a Confederate force. Powell, attempting to stem the Confederate tide, was hit above the right elbow by a minié ball, a thumb-sized lead slug weighing about an ounce. It shattered his arm, forcing an amputation. After weeks spent in a hospital recovering, Powell could have been mustered out of the army with honor; instead, he chose to rejoin Grant for the siege of Vicksburg and later William Tecumseh Sherman for the famous March to the Sea. His injury finally forced him to leave active service in January 1865.

But Powell did not sit on his porch and tell war stories. A restless spirit, he began teaching courses in geology at Illinois Wesleyan College. But he wanted his students to experience more than just classroom lectures, so he organized field trips into the western states and territories. These were epics in their own right; crossing the plains during the height of the Cheyenne wars, when stage stations were being burned and immigrants murdered, he led his students into the mountains of Colorado. There they climbed Pikes Peak and Longs Peak and explored the wild country west of the Rocky Mountains, still the home of the as-yet-untamed Utes. It was during this time that Powell conceived the idea of exploring the canyons of the Green and Colorado Rivers, to solve the "last great geographic problem" in the Lower 48.

The epic story of Powell's exploration of the Green and Colorado Rivers from May to August 1869 has been so well and often told that there is really no need to go into it at any length again; the fact that you have this volume in your hands suggests a familiarity with the basics of the journey. Suffice to say that Powell and his ever-dwindling crew labored, cursed, and grumbled their way down eleven hundred miles of remote, uncharted canyons, lining most of the rapids, running them only when they had to. The men worried about their shrinking supplies while Powell, as seemingly oblivious to the hardship and hunger as he was to the pain in his mangled arm, "geologized," collected fossils, climbed the cliffs for readings, and waxed rhapsodic in his notes about the beauties of the canyons, the music of the waters, and the glories of the dawn colors. All the while, though, the fear of the unknown chewed at the men's spirits as the lack of food gnawed their insides. Finally even Powell had to admit that because of incipient starvation they did not have time to do the complete scientific survey that he had envisioned, and toward the end they were forced to hurry lest they all perish from hunger.

Nearing the end of the Grand Canyon, they came to a huge, apparently unrunnable rapid, worse than anything they had encountered before. With success within their grasp but with uncertainty raging and frothing below their camp, three of Powell's men—the brothers O. W. and Seneca Howland and Bill Dunn—decided they had had enough, telling Powell they were abandoning the river and would strike out overland, to try to reach the Mormon settlements to the west. From readings with sextant and chronometer, Powell knew they were close to the end of the Grand Canyon, but what lay between was the crux. Even if they made it through the treacherous rapid below them, was there a huge, impassable waterfall just beyond that would leave them stranded and starving? Powell polled the other men and they all agreed to keep to the river. In a poignant morning scene, the two groups lined up to divide the pitiful last rations and exchange good wishes, each thinking the other was taking the more dangerous course. Abandoning the *Emma Dean*, the most worn out of the boats (another had been lost at the start of the journey, in Disaster Falls in the Canyon of Lodore, so they were now down to two boats) the three who were hiking out helped the six remaining river voyagers reload the boats and position them for the run. To their surprise, the rapid was easier than it appeared and they shot through, scarcely shipping a bucket of water. They pulled over below and waved and fired off their guns, to get the atten-

tion of the trio that had begun their climb out of the canyon, but there was no response and finally they had to go on. After coming to what really was the worst rapid they had encountered—Lava Cliff, now under Lake Mead—and somehow navigating it, the river travelers debouched from the Grand Canyon on August 30, 1869.

Meeting a local Mormon and his sons, who had been sent by Brigham Young to watch for wreckage and bodies of the Powell expedition, Powell and his brother, Walter Henry Powell, who had served on the crew, began a triumphal trip back across the country to Washington, D.C., where John Wesley was lionized as an explorer and scientist. Even while he was basking in the adulation of the crowds and his colleagues, however, Powell knew that he hadn't really accomplished much true scientific work. The trip had turned into a race for survival at the end, and Powell knew he would have to go back and do it all again if he really wanted to accomplish what he had initially set out to do. Indeed, about the only scientific fact that was well established, in Otis "Dock" Marston's arch phrase, was that there were banks along both sides of the river sufficient to line a boat. Besides, Powell wanted to try to find out the fates of the three who left and had seemingly vanished into the wilderness; nothing had been heard of them since that fateful morning in the depths of the Grand Canyon, at the head of what came to be called Separation Rapid.

So, capitalizing on his new-found fame, in 1871 Powell was granted funding from the government to pay for another river survey, now titled the Geographical and Geological Survey of the Rocky Mountain Region. Boats, equipment, supplies, and instruments were to be the same—with one major exception—but this time there would be no cantankerous frontiersmen to challenge his decisions. Powell recruited men with whom he had served in the Civil War, men with scientific backgrounds and training, and members of his own family. Almon Harris Thompson was all three; a veteran, a surveyor and map-maker, and Powell's brother-in-law. Others included Powell's young cousin, Walter Clement Powell; John Steward, whom Powell had met in the trenches around Vicksburg while they both were looking for fossils; and the topographer Francis Marion Bishop, who had been a student at Illinois Wesleyan University after active service in the Army of the Potomac. To ensure that they did not run short of supplies, Powell arranged for pack trains to meet them at certain places within the canyons with more flour, bacon, and beans. In the meantime, he had received the sad news that the three who had left the river had been killed by Indians on the north rim of the Grand Canyon; to prevent a repeat of that, Powell contacted Jacob Hamblin, a Mormon scout in southern Utah who served as Brigham Young's ambassador to the tribes, to assure the Indians that Powell's men meant them no harm.

The important change in equipment was the addition of a camera. During the Civil War, pioneers in photography like Matthew Brady, Alexander Gardner, and Timothy O'Sullivan had proven that despite the cumbersome wet-plate process with its jars of chemicals and fragile glass plates, cameras could be used in the field. Powell was quick to realize that here was a way to document his survey and the scientific and scenic wonders he knew were there, and publish them to the world. His contemporaries, Ferdinand V. Hayden and Lt. George Wheeler, were including photographers in their expeditions, so Powell knew it could be done. To that end, he hired E. O. Beaman, a landscape photographer from New York. Beaman brought the most up-to-date camera of the time, which consisted of a big wooden box which held the lens and a light-proof holder for glass plates. To take a picture, the photographer had to first coat the glass plate with a collodion solution containing silver nitrate, under a dark hood. While the plate was still wet, it had to be inserted into the camera. The photographer then removed the lens cap for a certain period of time, exposing the plate to the light. It then had to be "fixed," or developed, right then or the image would fade. From the time of coating the plate with collodion, the photographer had about ten minutes to make the exposure and fix the image. So besides the camera, the photographer had to carry jars of the chemicals needed, racks of glass plates, and a big tripod, since exposures were so long the camera had to be absolutely still.

As can be imagined, doing this in the field was a great deal of work. The camera itself was bulky, the tripod heavy, the vats of noxious chemicals and boxes of glass plates fragile. Walter Clement Powell, brought along as the photographer's assistant, called it the "infernal mountain howitzer," and complained often in his journal about having to haul the whole thing up the side of a mountain for a shot of the scenery. But Beaman was used to this, and from the start of the expedition in Green River, Wyoming, in May 1871, took numerous views of the river and the canyons, their camps, and the rapids. Beaman took photos all the way down the river to the head end of Glen Canyon, where Powell ended the journey for that year. But Beaman and Powell, as it turned out, had differing views on the purpose of the journey: Beaman wanted to sell his photos; Powell told him no, and in January 1872, Beaman left to go off on his own. To replace him, Powell hired James Fennemore, a Salt Lake City photographer who was working for Charles Roscoe Savage. Fennemore accompanied Thompson to retrieve one of the boats, which had been cached in Glen Canyon, but it turned out his health was not up to the rigors of such arduous travel, and he was forced to resign.

The obvious choice for his replacement would seem to have been Powell's young cousin, Walter Clement (Clem); he had been helping first Beaman and then Fennemore for close to a year, but Clem could never get the hang of the delicate mixing of chemicals, nor how long to expose the plate, nor any of the necessary details. But as it turned out, Powell already had a natural photographer on the crew: Jack Hillers. Hillers, a German immigrant, had come to America in time to fight for the Union during the Civil War. He later served in western outposts and was discharged in Salt Lake City in 1870. Through a chance meeting, he was hired by Powell "to pull an oar" and as a general camp helper. Hillers began working first with Beaman and later with Fennemore, from whom he learned how to mix the collodion solution and make a photograph. Hillers took to photography quickly, and from that time on was an official photographer for Powell and later for the U.S. Geological Survey.

Even though John Wesley Powell was nominally in charge of the second expedition, the real leader was Almon Harris Thompson, the second-in-command. Powell was along only for short periods on the river survey; the rest of the time he was raising funds, or investigating the fate of the three crewmen from 1869, or tending to his pregnant wife, Emma, who had fallen ill in Salt Lake City while she waited for her husband to complete the river survey. Thompson was a good, no-nonsense leader, and though the second expedition had its share of adventures and close calls, there were, as Thompson exulted at the end of the Canyon of Lodore, "no disasters." Learning of his wife's illness, Powell left before the party entered Desolation Canyon, leaving Thompson in charge. Powell returned for the leg through Labyrinth, Stillwater, Cataract, and Glen Canyons, where he decided that the river survey would conclude for the year. One of the boats was buried in the sand in Glen Canyon, while the party continued to Lees Ferry at the mouth of the Paria River. During the winter of 1871 to 1872, while Powell was traveling and raising funds, the survey worked out of Kanab, Utah, compiling a map of the Kaibab Plateau.

When the river survey resumed later that summer, the crew was unchanged save that Hillers was now the official photographer. On high water, the expedition continued through the Grand Canyon to their first scheduled resupply at Kanab Creek, about halfway through the Grand Canyon. There the pack train brought not only new supplies, but news: the Shivwits, who had killed the three crewmen in 1869, were reportedly once again up in arms over depredations by miners, making any exit from the canyon other than down the river a dangerous proposition. Considering the high water, the lateness of the season—by now it was late September—and the dangerous rapids to come, Powell called the men together and said, "Well, boys, our journey's through." There were no objections; as Hillers wrote, "We felt like thanking God."

The Powell survey stayed around southern Utah for the next three years, until 1875, enlarging their maps, establishing points for more surveys, and conducting studies of geological features, including what Powell named the Grand Staircase, now a national monument. Powell, always curious, took his own studies in a new direction by beginning to compile lexicons of Native American languages and recording their lifestyles through a series of photographs taken by Jack Hillers. In the course of the survey, Powell brought in men who would become some of the most famous names in American arts and science, such as Grove Karl Gilbert—whose *Geology of the Henry Mountains* is a classic geology text to this day—and Clarence Dutton, who wrote the *Report on the High Plateaus of Utah* and whose *Tertiary History of the Grand Canyon*, with its fabulous detailed drawings of the geology of the canyon, is as much a work of art as it is of science. Powell's own *Report on the Lands of the Arid Region* was the opening round in a debate about the proper use of land in the arid West that still rages today. The original members of the 1871 Powell expedition made their own contributions as well. Frances Marion Bishop completed the first detailed map of the canyons of the Green and Colorado before settling in Salt Lake City, while Frederick S. Dellenbaugh, at seventeen the youngest in the 1871 crew, became the unofficial chronicler and apologist for Major Powell. Dellenbaugh carried on lengthy correspondences with anyone who had anything to do with the Colorado River, testified in the Colorado Riverbed Case in the late 1920s, and produced two books about his experiences: *A Canyon Voyage: The Narrative of the Second Powell Expedition Down the Green-Colorado River from Wyoming, and the Explorations on Land, in the Years 1871 and 1872*, the story of the second expedition (because Powell, for reasons he took to his grave, ignored the 1871 crew in his official report), and *The romance of the Colorado river; the story of its discovery in 1540, with an account of the later explorations, and with special reference to the voyages of Powell through the line of the great canyons*. Dellenbaugh also set himself up as Powell's defender, because later in his career and even after his death in 1902, Powell had many critics.

Today the Powell expeditions are perhaps the best known of all explorations of North America; library shelves are lined with books about the subject of all kinds: biographies, historiographic essays, novels, children's books, reinterpretations. In addition, articles have been written in scholarly journals and popular magazines, stamps have been issued, documentaries and a feature film produced, poems and songs composed, all commemorating his epic journeys. And yet, for all that literature, for all the thousands of words written by and about Powell and his men, until relatively recently the core documents of the story—the actual journals and other writings of the members of his two expeditions—were rare books, to be found only in antiquarian bookshops or through dealers in precious and rare volumes. The writings of seventeen of the members of the two crews were gathered by the Utah State Historical Society and published from 1947 through 1949 in the *Utah Historical Quarterly* as volumes 15 through 17. These were widely hailed as a great boon to scholars and historians, but they soon fell out of print and within a few years were very hard to find. By the beginning of the twenty-first century, they were commanding prices of close to three hundred dollars each, a hefty sum by anyone's reckoning. Then, to the great joy of Powell scholars and Colorado River historians, starting in 2009, the University of Utah Press and the Utah State Historical Society cooperated in reprinting volumes 15 through 17 of the *Utah Historical Quarterly* in paperback (along with Almon Harris Thompson's journal, which had been published as volume 7 of the *Utah Historical Quarterly* in 1939). Now, for the first time in years, these foundational documents, essential to any study of Powell and his achievements, were easily and inexpensively available.

But there were still a few of the original Powell documents that had likewise gone out of print and thus were not easily obtained by today's readers: *"Photographed All the Best Scenery": Jack Hillers's Diary of the Powell Expeditions, 1871–1875*, published by the University of Utah Press in 1972; "John Wesley Powell's Journal: Colorado River Exploration, 1871–72," published in the *Smithsonian Journal of History* (volume 3, Summer 1968); "The Lost Journal of John Colton Sumner," from volume 37 of the *Utah Historical Quarterly*, published in 1969 (the "Powell Centennial" edition); "F. S. Dellenbaugh of the Colorado: Some Letters Pertaining to the Powell Voyages and the History of the Colorado River," published in the same issue of the Utah Historical Quarterly; and Frances Marion Bishop's maps, also published in that issue of the *Utah Historical Quarterly*. Now, to round out their goal of publishing all of the original Powell documents, and to further their service to historians, scholars, and the general public, the University of Utah Press has gathered these disparate documents together in the present volume. With the publication of *Cleaving an Unknown World,* thus completing the reprinting of all the original sources on Powell, readers can once again enjoy Major Powell's florid prose, Jack Sumner's ironic wit, Dellenbaugh's wide-eyed, youthful wonder, and Jack Hillers's clear impressions of the canyons of the Green and Colorado when they were still wild places. The University of Utah Press, the Utah State Historical Society, and all the other publishers who cooperated with this project are to be congratulated for this work, and have earned the gratitude of a whole new generation of readers, historians, and river runners.

Notes
1. *Report of the Exploring Expedition to the Rocky Mountains in the year 1842 and to Oregon and North California in the Years 1843–44*, by John Charles Fremont (Washington, D.C.: Gales and Seaton, 1845), 129–30.

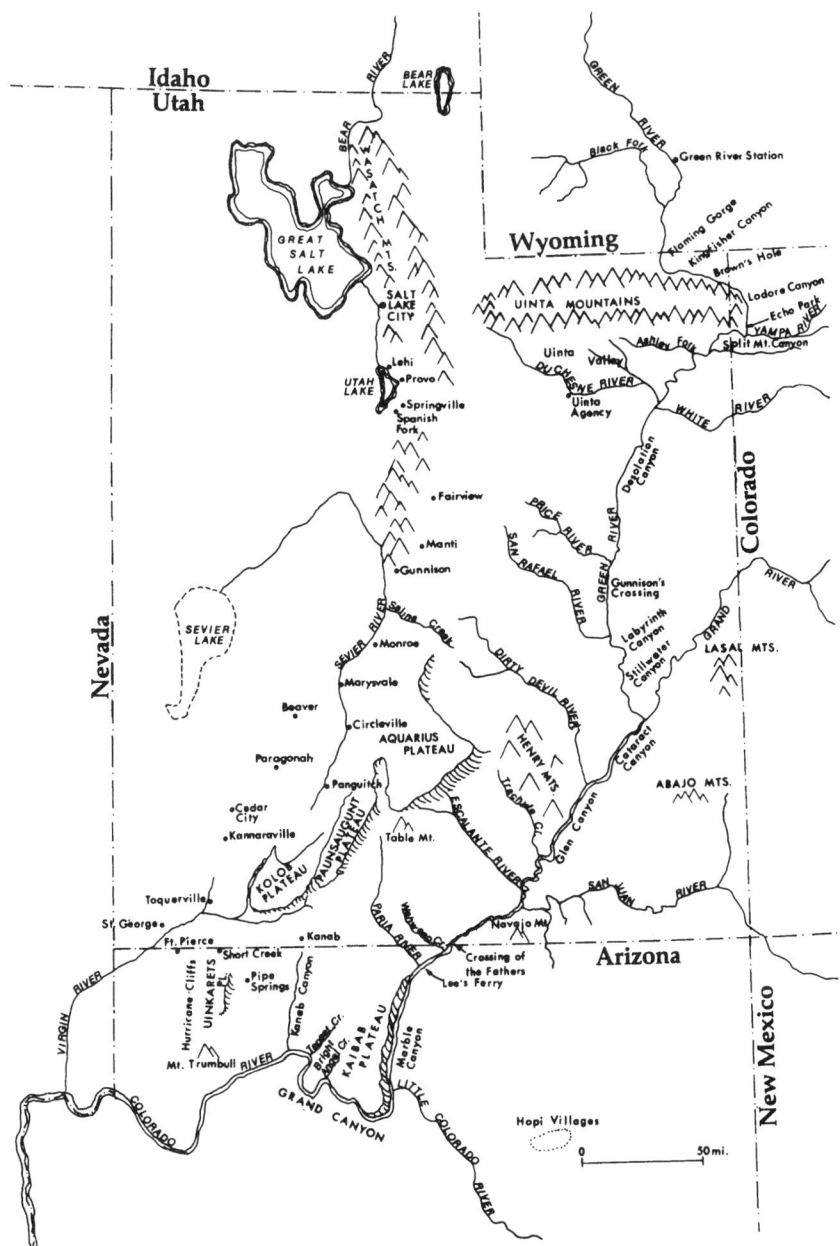

A Map of the Green and Colorado Rivers 1871–1872

The Members of the Powell Expedition, 1871–1872

John Wesley Powell (1834–1902). Powell was born in New York. In the Civil War he achieved the rank of major, a title he retained throughout his life. He lost his lower right arm at the battle of Shiloh. After the war he became a professor, leading natural history expeditions to the Rocky Mountains in 1867–68, and in 1869 leading a party on the first daring descent of the Green and Colorado Rivers. In 1879 he became the first Director of the Bureau of (American) Ethnology of the Smithsonian Institution. From 1881 to 1894 he was also Director of the United States Geological Survey. Powell died in 1902, internationally known as a scientist, administrator, and conservationist.

Almon Harris Thompson (1839–1906). Thompson was called Prof. Born in New Hampshire, Thompson moved with his family to Illinois where he graduated from Wheaton College in 1861. He married Ellen Powell, J. W. Powell's sister. Thompson joined J. W. Powell on the latter's reconnaissance of southern Utah and Arizona in 1870 in preparation for the second river expedition. Thompson became the chief topographer of the expedition and the de facto field supervisor. He was the chief geographer of the United States Geological Survey from 1879 until his death in 1906. The map of the Colorado River area was largely Thompson's work.

Frederick Samuel Dellenbaugh (1853–1935). Dellenbaugh was nicknamed Fred, Rusty, and Sandy by members of the expedition. He was a distant relative of Almon Harris Thompson and was only seventeen when he hired on as an artist for the Powell expedition. In 1873 he left the expedition and traveled in the West and Europe. Later he joined the Harriman expedition to Alaska and Siberia and made trips to South America and the Caribbean. He was a founder of the Explorers Club, wrote two books on the Powell expedition, and was responsible for collecting and preserving many of the diaries of members of the Powell expedition.

John K. (Jack) Hillers (1843–1925). Jack Hillers was born in Hanover, Germany, and came to the United States in 1852. He entered the Union Army during the Civil War and remained until 1870. In 1871 he chanced to meet John Wesley Powell in Salt Lake City and signed on as a boatman for Powell's second expedition down the Colorado River. Hillers later became Powell's chief photographer, serving in that capacity until 1881 when he became chief photographer of the United States Geological Survey. He held that post until 1900 when he retired. He continued on a part-time basis until 1919. He died in 1925 and was buried in Arlington National Cemetery.

Francis Marion Bishop (1843–1933). Bishop was nicknamed Cap and Bish. He was born in New York. Running away from home at age seventeen to join the Union Army, Bishop was promoted to first lieutenant for bravery at the first battle of Bull Run and at Antietam. He was badly wounded at Fredericksburg in 1862 but was eventually promoted to captain. Bishop entered Illinois Wesleyan University where Powell hired him as topographer for the second expedition. Bishop in 1872 settled in Salt Lake City. He taught at the University of Deseret and in 1877 opened an assay office. In 1873 he married Alzina Pratt, a daughter of the Mormon leader Orson Pratt.

Walter Clement Powell (1850–1883). Nicknamed Clem by the expedition members, Powell was J. W. Powell's first cousin. His parents died when he was six and he was raised by J. W. Powell's family. Clem was supposed to learn photography and become a permanent member of the expedition, but he never mastered the art and was replaced by Hillers. Clem returned to Illinois and entered the pharmacy business. He later married Mary Breasted, the sister of the famous orientalist and historian James Henry Breasted. He moved to Omaha, Nebraska, and became a prominent wholesale druggist. He died suddenly in 1883.

John F. Steward (1841–1915). Steward was nicknamed Ford, Sergeant, and Truthful James by members of the expedition. He was born in Illinois. Steward joined the Union Army in 1862 and in 1864 met J. W. Powell; the two men collected fossils in the trenches around Vicksburg during the long siege of that city. Powell later hired him for the second river expedition. Many of the geological sections in Powell's later works on Colorado River geology are credited to Steward. The exertions of the expedition exacerbated Steward's war wounds and he returned to Illinois in 1872. He later became an executive of the International Harvester Company.

E. O. Beaman. Little is known about Beaman's life. He apparently began working as a professional photographer after the Civil War. Powell hired him as a photographer on the recommendation of the E. and M. T. Anthony Photographic Supply Company of New York. He remained with the expedition until January 1872 when he and Powell had a disagreement and Beaman resigned. He subsequently made a trip to the Hopi mesas and published articles on the river trip and his Hopi trip in Appleton's Journal.

Andrew J. Hattan (1841–1919). Hattan was nicknamed Andy and General. Hattan first met Powell during the Civil War. He later joined the expedition as cook and boatman. In 1872 he returned to his home in Ohio where he worked as a farmer and teamster for the rest of his life.

Stephen Vandiver Jones (1840–1920). Jones was called Deacon by members of the expedition. Jones was born in Wisconsin and later moved with his family to Illinois. In 1870 he became principal of the Washburn, Illinois, school where he met Almon Harris Thompson, Powell's brother-in-law. Through Thompson, Jones was hired as an assistant topographer. He left the expedition in late 1872 and returned to his teaching duties in Illinois. In 1883 he moved to Dakota Territory where he became an outstanding attorney, helping to frame the constitution and legal code for South Dakota. He remained in that state until his death in 1920.

James Fennemore (1849–1941). Fennemore was born in London and presumably came to Utah as a Mormon convert. He was working in Charles R. Savage's photographic gallery in Salt Lake City when he was hired by Powell to replace Beaman. Early in 1872 he joined the expedition but ill health forced him to resign prior to the 1872 trip through Marble and Grand canyons. His most important contribution was teaching Hillers photography. Fennemore later opened his own photographic gallery. He is known for his portraits of Brigham Young and of John D. Lee seated on his own coffin prior to his execution for his alleged role in the Mountain Meadows massacre.

Frank Richardson. Richardson was called Little Breeches by the expedition members. Very little is known about this man. Powell sent him home once the river party reached Brown's Park, apparently because he was not strong enough for the trip.

PART 1

John Wesley Powell's Journal

Colorado River Exploration, 1871–1872

John Wesley Powell

Introduction by Don D. Fowler And Catherine S. Fowler

On 22 May 1871, John Wesley Powell and ten men set out in three specially built boats from Green River Station in Wyoming Territory. Their aim was to follow the Green to its confluence with the Colorado River and continue beyond the Grand Canyon to the Virgin River mouth. Their purpose was to gather geological and topographical data on this largely unknown region.

Two years earlier, on 24 May 1869, Powell and nine men had started in four boats from the same place with the same aim and purpose. Early in the trip a boat had been lost at "Disaster Falls." One boat was abandoned at Separation Rapids where three men left the party, fearing death by drowning in swift Grand Canyon waters. They climbed out the north canyon rim and were killed by Shivwits Indians. There were reports that the entire party had been drowned. When the survivors arrived at the mouth of the Virgin River—on 30 August 1869, after a fourteen-week passage—they found a search party fishing for their remains.

Scientific data was collected on this first trip, but not enough, and some records were lost. Primarily, Powell had learned that the Green and Colorado were navigable, and he gained a good general impression of the region. He felt that a second trip was necessary to obtain astronomic and hypsometric observations on which to base accurate topographic maps, and to make a thorough study of the area's structural geology and physiography.

As a youth, Powell had developed a great interest in natural history. After sporadic formal education and a stint of teaching in Illinois, he enlisted in the Union Army in 1861 as a private. He emerged from the Civil War (after losing an arm at Shiloh) with the rank of major. Having decided to continue his career as naturalist, geologist, and ethnologist, he became Professor of Geology at Illinois Wesleyan University and soon moved across town to Illinois Normal University. In 1867 he led a party of friends and students on a natural history expedition into the Rocky Mountains. During a second trip in 1868, he spent the winter on the White River in northwest Colorado, camping at a spot now called Powell Bottoms. During that winter he explored much of the Green River country, and determined to explore both the Green and Colorado Rivers.

In the Green River and its canyons he envisaged the difficulty of running these waters, and foresaw the supply problems for such a trip. Early in 1869 he returned to Illinois to enlist backing for his expedition. He gained the support of Illinois Normal, the Illinois Natural History Society, several private individuals, and the loan of some instruments from the Smithsonian Institution. He also put up a large sum of his own money.

Powell had special boats built in Chicago. They were hauled to Green River Station on the just-completed transcontinental railroad. He brought ten months of supplies against the possibility of being caught by winter ice and having to spend several months somewhere in the depths of the canyons, a wise precaution which turned out to be unnecessary.

The experience gained in 1869 led Powell to improve his plans for the second trip. He engaged Jacob Hamblin—the Mormon pioneer, explorer, and Indian missionary—to bring supplies to certain accessible places along the river. Powell hoped that Hamblin's party would find an overland route to the mouth of the Dirty Devil River. Their failure to do so caused some changes in plans, as will be seen. (A route to the mouth of the Dirty Devil was not discovered until 1872, and then by Powell's own men.)

After the first trip, Powell went to Washington to gain congressional support for further exploration, hoping to develop a survey similar to those of Wheeler, King, and Hayden, already in operation. On 12 July 1870 the United States Congress, by an appropriation of $10,000, created the "Geographical and Geological Survey of the Rocky Mountain Region, J. W. Powell in Charge," and assigned the project to the Smithsonian Institution.[1]

Powell immediately returned to the canyon country to prepare for the second river trip. In the fall of 1870 he went with Jacob Hamblin onto Kaibab and Uinkarets Plateaus in northern Arizona to make peace with the Indians (he wanted no repetition of the 1869 incident) and to scout out supply routes to the river.

Powell hired ten men to make the second trip. Almon H. Thompson—Powell's brother-in-law and a school teacher who had been with him in 1867—was in charge of topographic work. E. O. Beaman, of New York, was the photographer. John F. Steward, with whom Powell had collected fossil shells in the trenches before Vicksburg, was the geologist. Frederick Samuel Dellenbaugh—seventeen years old and distantly related to Thompson—was to be artist and boatman. Thompson's topographical assistants were Captain Francis Marion Bishop, a Union Army veteran, and Stephen Vandiver Jones. Both men were friends of Thompson and were acquainted with Powell. The latter's first cousin, Walter Clement Powell, was assistant to Beaman. Andrew J. Hattan—cook and boatman—had met Powell in the army. Another boatman, John K. Hillers (called "Bismark" by his companions) whom Powell met in Salt Lake City, later became a photographer and remained associated with Powell until 1900. The tenth man, Frank Richardson, left the party early in the trip at Brown's Hole to return home. Of the ten, only Beaman was professional; the rest were interested amateurs.

For the second trip Powell had three boats built, similar in design to those used on the first trip, but with watertight compartments at the ends and in the middle for buoyancy. The boats were named *Emma Dean*, for Powell's wife, *Nellie Powell*, for Mrs. Thompson, and *Cañonita*.

In 1875, Powell published his famed *Explorations of the Colorado River of the West and Its Tributaries....*[2] Part 2 of this volume—"On the Physical Features of the Valley of the Colorado"—is a remarkably clear and concise discussion of the geology and physiography of the canyon country. It was a pioneering work in terms of the area studied and in a number of its concepts.

The first part of this work is a compilation of Powell's and

Thompson's journals, some newspaper articles Powell had written during the first trip, and some articles subsequently written for *Scribner's Magazine*. This "Part First" is written as if it were a narrative of the 1869 river trip. In fact, however, Powell includes incidents and observations made during both the 1869 and 1871 trips (as well as the continuation of the 1872 trip through the Grand Canyon), together with the exploration of Kanab Canyon and the Parunaweap Canyon–Zion Park area in the fall of 1871. All incidents are dated 1869. There is no mention of the members of the second river trip, nor even that there was a second trip.

This "telescoping" of events has exercised the indignation of historians and others for many years. Clearly, such a document is not a historical narrative. It is, however, a narrative of adventure and is written in such a tone. One must agree with Powell's biographers[3] that his probable purpose in writing "Part First" as he did was to excite the attention of the public and the Congress, thereby creating a friendlier atmosphere for continued support of his intended programs. In view of all his subsequent accomplishments as geologist, anthropologist, conservationist, and administrator of scientific endeavors, Powell can perhaps be forgiven this foray into the realm of public relations.

We have not attempted to reconcile the journal printed here with Powell's *Explorations*, that is, to sort out which incidents related as happening in 1869 actually occurred on the second trip, as, for example, the naming of the "Cliff of the Harp." Our purpose, rather, is to complete the historical record of Powell's second river trip.

Powell's survey continued during the 1870s, at first under Smithsonian auspices and later under the Department of the Interior. During these years Powell personally conducted ethnographic and geological research in Utah, Arizona, and Wyoming, directing as well the geological and topographic work of others.

His extensive collections of Indian artifacts are now in the United States National Museum.

Powell's own geological studies and those made under his direction have long been regarded as classics in the field.[4] Very little of his extensive notes and manuscripts on the Indians have been published, but they are now being edited.[5]

The Journal of the Second Trip

The Powell expeditions are well documented. The 1875 *Explorations* were later revised, rearranged, and expanded by Powell and published as *Canyons of the Colorado* in 1895.[6] In 1908 Frederick Dellenbaugh published *A Canyon Voyage*, an expansion of his own journal.[7] In 1939 Thompson's journal was published by the Utah Historical Society.[8] Between 1947 and 1949 the Society also published most of the extant diaries and journals from both the 1869 and the 1871–1872 trips together with some newspaper articles written by various members of the two expeditions and biographical data on the men themselves.[9]

Powell's own journal of the second trip, however, was not published because—according to the editors—of its fragmentary nature. This fragmentation was owing, in part, to Powell's being away from the river for extended periods during the second trip. During these times he did not keep a regular journal. Another probable reason for its fragmentation is that a section of the journal was misplaced in the Bureau of American Ethnology archives, and hence was not seen by the Utah Historical Society editors. This inference is strengthened by the fact that excerpts from Powell's journal of the second trip were printed as footnotes to the other journals, but no excerpts appear from the misplaced section.

The Powell journals consist of a set of notebooks with board covers containing penciled notes. Some are geological notes from the two river trips and the overland trip of 1870; others are ethnographic notes on the Southern Paiute Indians; the remainder constitute the river-trip journals; one volume on the 1869 trip, a second on the 1871 trip, and the third on the 1872 trip through the Grand Canyon. The "missing" section, which belongs in the second volume, contains some miscellaneous ethnographic notes on the Uintah Utes. These notes were made while Powell was traveling from the Uintah Indian agency to the headwaters of the Sevier River in late July 1871.[10] But the main body of material in this section is Powell's 1871 journal for the period 11 June to 7 July.

When the Bureau of American Ethnology archives (now incorporated into the Smithsonian Office of Anthropology archives) were reorganized in 1926, the 11 June–7 July section of Powell's journal seems to have been placed with his other ethnographic materials. Subsequently, they were apparently overlooked by the editors of the *Utah Historical Quarterly* series.

What is printed here, then, is Powell's complete journal for the time he was actually on the river during the 1871 and 1872 trips. The ethnographic materials on the Uintah Utes are omitted; they will appear in the Powell ethnographic manuscripts.[11]

In preparing the journal for publication we have corrected Powell's infrequent spelling errors, added punctuation as necessary to clarify meaning, indicated Powell's original rendering in notes at the bottom of the page, expanded his abbreviations, and occasionally inserted a word or letter in brackets. Since most of the mountains, canyons, buttes, cliffs, and other formations were being named for the first time by the party as it proceeded, we have attempted to indicate the sources of the names given. We have given, in brackets, the present name of a formation or feature if it differs from the name given by Powell.

The photographs herein are from the original wet-plate negatives made by Beaman during the 1871 segment of the trip and by Hillers during the 1872 segment, as well as later when Powell sent him overland into the Green River and Grand Canyon areas. Hillers learned photography from Beaman before the latter quit Powell's survey in late 1871.[12]

I. Journal from Green River to Brown's Park, 22 May–9 June 1871

May 22 1871, Monday[13]

Started from Green River Station at 10 a.m. Breakfasted with Mr. Fields.[14] The good people of the town came out to see us start and gave three cheers as we left the bank. We ran down against a stiff breeze to noon camp running aground on a bar just before stopping. After dinner we, in running around a bend to sight the *Nellie Powell* and *Cañonita*, ran very hard on rocky wall carried by current. Camped at cabin.[15] Found fossils, etc. Camp No. 2.

May 23, Tuesday

The rain on our faces waked us at daybreak this morning and it continued to rain and snow until 10 a.m. Started after dinner against strong wind. Passed cabin. I went in. Beaver heads and skeletons scattered about. About 3 [p.m.] saw two men, trappers, on the right near cliff. Sent fossils in box to Mr. Fields. Camped in cotton-wood grove. Bishop and Clement paced a distance ¾ mile. Beaman took photos. Thompson measured two cliffs. Camp No. 3.

May 24, Wednesday

Started early. Stopped at 9 a.m. Beaman took photos. Bishop took topography. Steward and I climbed to foot of Needle Butte. Found fossil fishes. Creek comes in from east. Had beautiful view of the Uintahs looking down the valley of the Green. The summits of these towers, cliffs, and needles at the general level of the country. Beaman took photo looking up valley of needles and towers from noon camp. (This photo not preserved.)[16] Camped at mouth of Black's Fork, No. 4.

May 25, Thursday

Run today with low round hills on either side and low cliffs near river occasionally. Broad quiet and deep river. Lovely ride. Come to Deer Island[17] and kill three. Camp on left bank in beautiful meadow. Camp No. 5.

May 26, Friday

Come down to the tilted rocks and go up into camp. Steward traces strata back 3 or 4 miles. Thompson, Bishop, and I climb ridge and take bearings. After dinner Beaman takes photos. Camp No. 6.

May 27, Saturday

Steward and I hunt fossils and trace strata. Thompson takes his boat down to Flaming Gorge. Beaman follows. I go later. About 3 o'clock Thompson and I go to take a look at Horse Shoe [Horseshoe] Cañon.[18] Camp No. 7—at Flaming Gorge.

May 28, Sunday

Steward looks for fossils among the carboniferous rocks until noon. He and I trace strata and collect fossils from camp in Flaming Gorge[19] to Henry's Fork near old cabin and return at night. Find Jurassic fossils.

May 29, Monday

Come down to lower end of Horseshoe Cañon. Our first rapid is here. After dinner came down to head of Kingfisher Cañon[20] and camped. Camp No. 8 on rocks.

After coming down to Camp No. 8 Thompson and I walked out on the mountain to see Kingfisher Creek. (Started deer.) Had a view of the mouth of the creek where its pure waters emptied into the turbid Green. From this point (the west cliff of Kingfisher Cañon) had a fine view of the Green winding around at the base of bright red cliffs forming a vast amphitheater and a long line of huge cliffs to the northwest with the wooded valley of the Kingfisher Creek and the crystal stream winding in the midst and the green domes of mountains with patches of cedar and pine on the south of the valley.

May 30, Tuesday

Today with Bismark [Hillers] I climbed mountain south of Kingfisher Creek. Had fine view of the mountains to southwest covered with snow and over the cliffs of Henry's Fork to B.B.B.B.[21] Quien Hornet[22] too was covered with snow. Slowly the clouds enveloped the mountains and we came down. Steward and Dellenbaugh were making section. Same Camp No. 8.

May 31, Wednesday

Ran down into Kingfisher Cañon. Stopped in park at mouth of creek—wounded [a] deer. Took views. As we enter the park we turn to left, and looking back the way seems closed. Bold rocks of grey sandstone tower on the right; on the left crags and rocky slopes with scattered cedars, piñons, and firs. A beautiful meadow valley with clumps of alder. The Kingfisher comes in on the right through a narrow cañon valley with steep walls, the valley itself filled with alders and willows completely hiding the creek. Then the creek emerging from its own cañon meanders across the little park, its banks fringed with willows. The river flows at the foot of the cliff on the left and is hedged by a border of willows from the meadow on the right. Looking down river the walls seem almost to close where the river turns to the left around Beehive Point.[23] And out through the cañon you can see the green and wooded slopes of distant mountains and a single snowbank like a long white cloud dropped from the skies and caught by the mountain ridge. After taking pictures we rounded Beehive Point and camped at head of Red Cañon.[24]

June 1, Thursday

Thompson and Jones climbed cliff to south, Bishop to northwest, Steward and Richardson took section in top of Devonian. Beaman made pictures of the deeply worn rocks. Below Beehive Point these are in nearly verticle [vertical] strata of the carboniferous just above (geologically the Devonian).

Beaman has also taken pictures of the head of Red Cañon looking down and looking out, up, and also of the crags in the upturned carboniferous near Beehive Point.

June 2, Friday

Down into Red Cañon. Thompson's boat is upset by running too near rock on right. Lost compass, sponge, and camp kettle. Passed two creeks. Beaman takes photos at noon camp. Run down to first portage and complete it and camp on the old ground. [Camp] No. 10.

June 3, Saturday

Captain Bishop and Jones climb wall to right and tell of long wooded grassy slopes back to snow-clad mountains and of streams running down from these mountains through beautiful valleys. Beaman has taken pictures of the cliffs on either side of river. The dip of the rocks is to the north for this reason. The cliffs on the north side of the river are more bold, often nearly vertical to top, 1,500 to 1,600 feet high. Sometimes lines of pines and firs are arranged on narrow shelves formed by the strata. Or trees will be grouped irregularly up gulches coming down straight to the river. At other places the gulch will come down obliquely to the face of the wall and have its group of trees. The slopes on the south side are more gentle, often hollowed out between projecting cliffs that seem like huge stairways built from the river to the regions above, built for the giants of the elder days but now half in ruins. Photos Nos. 16-17-18 and stereos 37-38-39-40-41-42 illustrate these facts.[25] Still Camp No. 10.

June 4, Sunday

Came down to point on left where we let down with ropes. Afterward passed two boats tied up, oars laid away. These are the boats abandoned by the prospecting party two years ago.[26] Camped on island for dinner and remained the day. Thompson and I climbed mountain. Had view of the Notch and Green River below Camp No. 6. Good view, too, of the Uintah and the grassy slopes down to the Cañon. These are dotted with beautiful pine groves and clumps of fir with meandering creeks coming down from the mountains into the Green. Sat up late and told stories of Army life.[27] Camp No. 11.

June 5, Monday

Ran down over rapid river to point where two creeks come in, and stopped for photos. Got good views on both creeks. Francis Creek first, Cactus [Creek][28] below (Beaman has these names reversed). Came on to Ashley's Falls and made portage.[29] Here Beaman made a number of good photos. Camp No. 12.

June 6, Tuesday

Came down to noon camp where we had beautiful view of run and amphitheater slopes. Beaman took pictures. River very swift. Grand ride from the falls to Red Cañon Park. Above the park at Snow Creek Beaman took photos. Camped in Red Cañon Park under two grand old pines. Camp No. 13.

June 7, Wednesday

Thompson, Steward, Bishop, Clement, and I climbed mountain which we named Mt. Lena[30] at Bishop's suggestion. Had a fine climb first along a little brook then up a point to right among the cacti and painted cups then along the cliffs of the cañon of Ashley's Creek, which we crossed. Then up the mountain over rocks, among rocks, around rocks, through fallen timber. At last on the summit. What a view! The Wind River mountains on the north, 150 miles away covered with snow, the Wasatch and Uintahs on the west. These peaks glittering in the silver sheen of everlasting ice. Quien Hornet, too, away to the northeast and mountains and valleys in dim perspective in every direction. Then the ridges of upturned formations standing in long lines curiously curved and freted [fretted] and painted in bright and glowing colors. The geological record of the ages since life began. 27,000 feet of wave-built rock turned up to one view. Whoever saw at one view so long and so grand a record? None before.

Then we descended to a creek, made a cup of tea, ate our luncheon, and returned to camp by sunset. The mountain was found to be 3,300 feet above camp.

June 8, Thursday

Came down to Christian Hillman's cabin in Brown's Park [Brown's Hole] where I had camped twice before. Once when Emma was with me. After dinner Thompson and I went out among the rocks. Boys at work in camp on notes, maps, packing fossils, photos, etc. to be sent to Green River City.[31] Camp No. 14.

June 9, Friday

Still at Camp No. 14 preparing to send out our extras, finishing up work, etc.

II. Journal from Brown's Park to Mouth of Uintah [Duchesne] River, 11 June–7 July 1871

June 11, 1871, Sunday[32]

Left Harrell's Camp and came down through Swallow Cañon.[33] Camped for dinner on the rocks near the foot of the cañon in the shade of the cliff while Beaman was taking pictures above. Then a long ride through a beautiful valley on a broad river to Camp No. 15.

June 12, Monday

Jones, Steward, and I climb on to the plateau on the north side of park.[34] Beautiful grassy plateau with long lines of ledges of rock in direction of the strike, the strata dipping to the north. Beamans [Beaman] takes pictures of camp. Thompson climbed mountain to south.

June 13, Tuesday

Out on the beautiful river again winding among groves of cottonwood with meadows here and there and mountains all about. We lashed our boats together and quietly floated down with the current while I read "The Lady of the Lake" aloud.

June 14, Wednesday

After dinner Thompson, Clement, Bismark [Hillers], and I climbed mountain on left of the gate of Lodore.[35] Was an hour and twenty-five minutes in getting up. Huge cracks in the rocks adjacent to the wall were noticed. These are caused by the weight of the rocks.

June 15, Thursday

Jones and I climbed Mt. Cullow.[36] First over the river cliff, then through the cedars to old Indian trail, then along this trail up to the divide. Then along divide to summit. Hot! Dry! Hard work! But grand view!! Cañons shape and distribution of the peaks due to the strike and dip of the rocks.

June 16, Friday

Made map of Brown's Park.

June 17. Saturday

Waited until now for letters[37] then out into Lodore. Beaman stayed for views. Camped under box elders at "Winnie's Grotto."[38] Enjoyed greatly the scenery and read Longfellow by firelight. Was lulled to sleep at last by the roar of the falls below.

June 18. Sunday

Beaman took views of Winnie's Grotto, a stream of light pouring into the huge fissure.

Away back in the fissure is the grotto, an amphitheater in the rock with the water oozing out from a thin, soft stratum or seam and trickling down the sides covering the rocks in places with mosses and other water-loving cryptogams [cryptograms].

In the fissure gulch above are lodged some huge boulders. Looking out overhead from below you see an irregular, jaged [jagged], and crenulated ribbon of sky.

Then we ran down Disaster Falls.[39] Beaman took pictures of the falls from right bank and of the "Correl of Lodore." As we came up to the falls Thompson and I started a flock of sheep, eleven in number. They ran back into a huge hollow in the wall of the cañon and we found the signs that they had lived there for a long time (there were three lambs in the flock). So we gave the place the above name. Then we camped in the old ground among the cedars.[40]

June 19. Monday

Made the portages around falls. I had long walk along the river bank. Beaman took pictures of falls and of "Sage Gulch."

June 20. Tuesday

Loaded boats and made portage opposite Wreck Island,[41] dined, then ran down to mouth of Cascade Creek [Pot Creek?]. Then down to the same cedars on right bank where we camped [in 1869]. A huge vertical wall of rock rose from the opposite side of river. As I was lying on my blankets about 10 p.m. I saw Vega come over the wall. And so high was the cliff that the bright star seemed to come down into the cañon. Only a narrow strip of heavens with a few constellations could be seen. And the jaws of the cañon threatened to devour these. The cliff opposite was named the "Cliff of the Harp."[42] A peculiar point of rocks above was called the "Wheat Stack."

June 21. Wednesday

Pictures were taken this morning of all these points and we ran down to the old camp of 16 June 1869 at the head of Triplet Falls and the foot of Dunn's Cliff.[43] Camp [No.] 21.

June 22. Thursday

Beaman took views here. Thompson, Bisop [Bishop], and I climbed to the summit of Dunn's Cliff. Our way was up a deep gulch among pines and firs. Then along a high, ragged ledge of limestone to a point opposite Mt. Dawes. Away out here at the highest point of the ledge we found a monument that must have been erected many years ago.[44] After taking observations, etc., we descend to camp coming along the foot of the limestone ledge and lower shelving rock, then down the gulch to camp; a wild running descent. Camp [No.] 21.

June 23, Friday

Made portage around Triplet Falls and ran down to the head of Hell's Half Mile[45] and let down one boat. The upper portion of the fall is more precipitous and was called Boulder Falls, being filled with huge boulders that have been rolled down from gulch on the left. Here Beaman took pictures. Men very tired tonight. Camp No. 22.

June 24, Saturday

Thompson with the men let down the two other boats. I climbed the left wall along shelf on the ledge around to Alcove Brook. Started a sheep on the way that made a jump to gain the summit of a rock and fell off into the river, say about 40 or 50 feet. But got up and, struggling to shore, bounded away. On leaving Alcove Brook I started across the mountain and reached the head of Boulder Gulch up which the smoke was rolling in dense clouds. So I took my way farther down the river and had hard, weary climb to camp where I found the men just finishing the portage at one p.m. After dinner we ran down to the mouth of Leaping Brook [Rippling Brook] and camped.

Still the 24th

On our arrival at this camp Thompson, Beaman, and I go up the creek into a little domed alcove where the brook was rolling over mossy rocks. Then we climbed the craggs [crags] to left and had a fine view of the brook. One branch coming down from the right, another from the left. The right-hand branch was leaping over the rocks and making rippling mellow music in sweet soothing contrast to the roaring of the river, so mad and raging, that had been in our ears night and day for more than a week.

High, bold rocks of "transition sandstone" were on either side between the brooks, and back and above them all towered the craggs [crags] of creamy white limestone. So the dark red rocks below, the variagated [variegated] rocks at mid height, and the rich white rocks above with firs and pines and cedars growing from every fissure and shelf and decking the landscape with evergreen spray. And now and then a cluster of box elders on the margin of the brook where it would run for a few feet between two precipices on a level rock with spray of golden green. All these gave such a wealth of colors as could be equaled only by a firmament of storm clouds illumined by a midsummer's sunset.

June 25, Sunday

Beaman took views. After dinner let boats down with lines and ran to Alcove Brook, took pictures, and then ran into Echo Park[46] [Pat's Hole].

June 26, Monday

In camp at Echo Park. After dinner I go across the Yampa and up Dry Gulch, round to the Triassic Patches and down the creek.

June 27, Tuesday

Took the *Emma Dean* and with Jones, Bismark [Hillers], Beaman, and Hattan started up Bear [Yampa] River.

Killed rattlesnake. Camped for noon on right bank in shadow of cliff. Crossed after dinner. While Beaman took pictures, I climbed up the Ribbon beds and collected fossils. Then towed up to Grizzly Park,[47] camped for the night.

June 28, Wednesday

Jones and I climbed to cragg [crag]. After dinner towed up a bad rapid but not quite so bad as the one made the previous afternoon. Then on, rowing and towing through a beautiful cañon. Camped under the box elders [at Box Elder Park?] on the left bank (sunny afternoon, stopped for views).

June 29, Thursday

Started early. Stopped at 9 a.m. for views and got one on top of cañon. After noon rowed and towed up to Hattans Park and found (on examination) that we were nearly out of rations. Bud on the grass.

A long walk after supper. Saw a deer on the rock above me.

June 30, Friday

Climbed mountain starting early with Jones. Beaman takes the boat below and takes views. I return by noon; dine on baked fish. Start for Echo Park. Let down with lines three times on the way and arrive at 5 p.m. The scenery of this Yampa Cañon is very beautiful. Many curves in the river. On the outside of the cañon the walls are nearly or quite verticle [vertical] down to within 200 or 300 feet of the water and then usually slope. This slope is set with pines and firs growing out of crevasses or on benches. The face of the wall is very regular and beautifully curved. The points on the inside of the bends are usually craggy. Some beautiful gulches come in on both sides, both deep alcove and wide gulches.

Found a large spring in Grizzly Gulch.

July 1, Saturday

I stay in camp at Echo Park all day making map.

July 2, Sunday

Steward and I climb along the ledge to short fold to see if any "fault" can be observed.

July 3, Monday

Start from Echo Park. Beaman remains behind to take views. Run down through

Whirlpool Cañon way below through "Old Red Sandstone." Walls close up to water's edge verticle [vertical] for many hundred feet and then craggs [crags] reaching back. The cañon opens out below the falls. Camped at mouth of Brush Creek.[48]

July 4. Tuesday

Remained in camp all day. Thompson and others climb mountain. I sit on the bank of this beautiful creek and read. Fred [Dellenbaugh] "gets up" a Fourth of July supper.

July 5. Wednesday

Down through Whirlpool Cañon into Island Park at the entrance to Craggy Cañon.

July 6. Thursday

Collect fossils.

July 7. Friday

[On 7 July Powell, with Bishop and Jones, took the *Emma Dean* to the mouth of the Uintah (Duchesne) River; Thompson followed a few days later. From there Powell expected to go to the Uintah Indian Agency upstream for supplies, then return to the river. But, learning of his wife's illness (she was six months pregnant), he proceeded to Salt Lake City to see her, leaving instructions for Thompson to wait for him. Thompson left the crew camped at the river and went to the agency. On 24 July Powell returned to the agency to report that the attempts by Hamblin's men to reach the mouth of the Dirty Devil River had failed. Powell proposed to go south with Hamblin's men to try again to find the mouth of the Dirty Devil. Thompson was to continue down the Green to Gunnison's Crossing (about 5 miles north of present-day Green River, Utah) and wait for Powell there. Thompson was unhappy with this arrangement, but went. Powell, having failed to find a way into the Dirty Devil, procured a few supplies at Manti, Utah, and rejoined the party at the crossing on 29 August.

It remained for Thompson, in the spring of 1872, to find an overland route to the mouth of the Dirty Devil.[49]]

III. Journal from Gunnison's Butte to Crossing of the Fathers, 2 September–8 October 1871

Left two boats, the *Nell* and *Cañonita*, in Island Park and party in charge of Thompson on July 7.

Arrived on Green River at foot of Gunnison's Butte August 29, 1871 and joined party again.

Remained in camp 30 and 31 [August] and September 1, crossing river to new camp on the 1st.

September 2, 1871, Saturday

Ran down to Black Bluffs past aragonite beds.

September 3, Sunday

Jones and I walk out to point near the upturned rock which forms the plateau to the east between this valley and the cretaceous valley at foot of the Wasatch cliffs. Vast fields of naked earth seen. Sometimes almost as light as a bed of ashes. Hard walk home to camp. Steward and Bishop discover cave. Thompson takes over.

September 4, Monday

Run down to San Rafael through beds of purple and grey sandstone.

September 5, Tuesday

Jones and I go up the San Rafael and climb a conical butte. See Jones Peaks to west, a line of buttes south. Still beyond, volcanic mountains, the Sierra la Sal to east. The Blue Book Cliffs to north with the Brown Cliffs surmounting them.[50] Camped on the little river.

September 6, Wednesday

Returned to camp. Collected flint chips. Found many chips, a good number of unfinished arrowheads, and a few complete. Also found rocks which showed evidence of having been used for hammers and anvils. The chips were found in a space of 10 or 12 acres where old camps had been. Back on the hills the rubble stones had been much broken up, probably to obtain the material of which the arrowheads were made. Bushels of the chips could be collected. Bishop and Clement went out this day to take observations from Bishop's Butte. Did not return. We kept up fire during the night.

September 7, Thursday

At daybreak Hillers and I start out to hunt the lost men. We meet them four miles from camp coming slowly, being very tired. Then we go down to the cliff overhanging river and wait for boats. Storm comes on. River of red mud. After storm run down to Trin Alcoves[51] [Three Canyons]. Spread our canvas in a storm (Dellenbaugh's Butte).[52]

September 8, Friday

Climbed naked rocks in front of bend.[53] In every direction as far as we have seen for the past two or three days naked rocks prevail. Buttes are seen scattered on the landscape, now rounded into cones, now buttressed and columned, now carved out with alcoves and sunken recesses and pockets. All varying from orange to dark brown, often stained black. Curious arches, too, are seen on the vertical walls of the cañon. From the rounded rocks of this point with pockets filled by yesterday's rain we look off on a fine stretch of river and over the naked rocks and buttes, to the Blue Cliffs [Book Cliffs] and the Brown [or Roan] Cliffs beyond and above with cumuli piled over all.

After dinner Thompson and I explore the recesses of Trin-Alcove [Trin Alcove], the left one is an amphitheater turning spirally to left and up with overhanging shelves and then turning as you look up to the right where there is a series of water basins. From these the water comes down into the basin at the bottom of the cove. Huge rocks lie piled up below, and on the right the rocks arch overhead. The middle cove is a beautiful glen with verdure spread and trees scattered here and there. The right cove is a narrow winding gorge often with overhanging walls and mighty domes almost shutting out the light. (See pictures.)[54] This cañon had many oaks along the base of the walls.

September 9, Saturday

Came down to Bowknot Bend.[55] Camped at the foot of Broken Wall. Thompson and I climbed. Saw beautiful rainbows spanning the cañon during afternoon while on bar. The lower and plainer was quite apparent against the rocks 200 yards away.

September 10, Sunday

Beaman took pictures while boats went round the eastern loop [of Bowknot Bend]. During the afternoon I hunted fossils—Thompson climbed.

September 11, Monday

Still down, down, winding among the bends of Labirynth [Labyrinth] Cañon. Took views at noon on looking up long stretch of river, another looking across at Tower Cliffs [Horsethief Point?]. Then down through Tower Park [Woodruff and Mineral Bottoms?]. Took picture just before storm. Camped in Labirynth [Labyrinth] Park [Saddlehorse Bottom?] at foot of cañon.

September 12, Tuesday

Came on to head of Stillwater Cañon. Climbed out and took pictures, then run down (after dinner) to Old River Bed.[56]

The view from bluff was very grand. The river winding its way through the rocks to the southeast with overhanging walls. Naked terraces back to Orange Cliffs.[57] The terraces with sharp salients and deep retreating angles. Orange and deep-red buttes standing on the terraces; columned above, buttressed and fluted below. Some set with pinnacles and towers. Beyond the buttes to the southeast the Rock Forest,[58] and beyond the "forrest" the mountains [Abajo or Blue Mountains]. Everywhere naked solid rock or naked talus. From noon camp, one of these buttes stood in front of another so as to seem one with it and present the shape of a huge cross, so we named it the [Butte of the] Cross. This cañon has many drooping willow trees along the river. After dinner ran down to Old River bed and took more pictures. Found curious plants hanging from the rock in the angle formed by the gray and red homogeneous beds as I found them two years before.[59] The Rock [Turk's Head?] standing in the center of the old curve is about ⅓ mile long. Have a beautiful view of the curve of river.

September 13, Wednesday

Ran down to camp under Salt Cliff.[60] The grey homogeneous bed caps the cañon walls, still overhanging. Until noon. Camp. Now and then we obtain good views of some point of country beyond the cañon—where we turn a bend and there is a good stretch of river before us. Noon camp is on some shaded shelves. After dinner run down a short distance and take pictures. Thompson and I have a climb on to a "wall point" but can find no way to get up to the Orange Cliffs as the homogeneous grey still caps the cliffs and is everywhere overhanging.

Angles, pinnacles, gulches, towers, cliffs, buttes, beautifully carved talus, highly colored rocks; a wild grand dissolution everywhere.

September 14, Thursday

Collected fossils during forenoon. Run down to old Shinumo town.[61] Thompson and I try to climb out but fail. Find ruined houses on a shelf at the angle of gulch and main cañon. Collect many arrow chips and fragments of pottery and corn cob. Beaman takes pictures of river.

September 15, Friday

Climbed out and run down to junction of Grand and Green.[62] Climbed by aid of roaps [ropes]. Found large water jar and willow splits for Potaz.[63]

Grand view of rocks above found where the Indians had placed poles against the rock for the purpose of climbing out. Found stairway to river from old Shinumo [Peublo] house.

September 16, Saturday

Climbed out with Jones, Beaman, etc.[64] Take pictures of the Green, of the Colorado, of the distant Sinav-Tuweap, and of the Rock forrest [forest] [Needles District]. These pinnacles are carved out of the upper carboniferous red and grey beds. Jones and I wandered for several hours among the rock-bound glens at the head of the lateral gulches down the wall of the Colorado.[65] Found pottery and arrow chips.

September 17, Sunday

Climbed out again to the Sinav-Tuweap. Found abundance of water in pockets. Beaman took nine pictures.

Thompson and I wander among the strange rocks of the Sinav Tu weap [Sinav Tuweap]. These rocks are carved out of the upper carboniferous. Are quite homogeneous in structure but have beds of grey and beds of bright red alternating. The columns and beautiful rocks are set about the borders of weird glens at the head of lateral gulches. The whole country is naked rock. How shall I describe the Sinav Tuweap?

September 18, Monday

Climbed out with Jones to summit of wall on east side of Grand. Passed the head of gulch that runs down into the Colorado below [the confluence] 3 or 4 miles[66] and found an Indian trail for horses, and campfires that were probably made last winter. No doubt but that horses can be taken down to the Colorado at this point. If the Grand should be explored in boats the party could return by this route to the Sierra la Sal or Sierra Abajo during a season when water or snow is abundant, or perhaps by careful exploration water may be found at any season.[67]

September 19, Tuesday

Ran 10 miles.

This cañon [Cataract] has many hackberry trees above the river. Half mile below noon camp on left the strata are thrust up at a dip of 15° above and increasing down to the verge of the river where they dip 80°, and a bed of carboniferous black clay shales are thrust up. In the region are many sulphur springs.

September 20, Wednesday

Climbed out to Rattle Snake Butte.

Steward and Bishop spend the day in making a map to exhibit the complexity of dip among the salients, etc. Thompson takes observations. Beaman makes two pictures.

Dellenbaugh and I have a hard climb up gulch on right. Find no water in the pockets and suffer from thirst. A wild descent into the cañon.

The atmosphere was very clear and we had a good view of the Orange Cliffs, the Sinav Tuweap, the buttes, the terraces, the pinnacles, the Sierra la Sal, the Sierra Abajo, and a mountain to southwest [the Henrys?]. No snow on the Sierra la Sal as there was two years ago.

Dellenbaugh killed rattlesnake on summit of butte, with ten rattles on his tail. But the snake was quite small, say 20 or 24 inches in length.

September 21, Thursday

Down we go among the rapids. Huge rocks have fallen from the walls, great angular blocks scattered down the talus and strewn along the channel. The walls too are very craggy. Beaman took a good picture of these crags this morning. The walls have been gradually increasing from the junction where they were 1,200 feet high to this point (noon camp) where they are 1,600. The west wall is very bold and grand, nearly vertical. The waters make roaring music at the foot of the cliffs, plunging over falls and whirling and foaming among the rocks. The men work with a will that seems wonderful. Here we have cataracts. Hard work.

September 22, Friday

Make but a mile and ⅜ today. Letting our boats over three cataracts by very hard labor. The walls are about 2,000 feet high on right and nearly vertical, but on left are broken and craggy. The scenery is very grand and the roaring of the mad waters is something awful. Stop to repair boats rather early. At the falls huge blocks of rock obstruct the channel causing chutes and whirlpools; and still the water tumbles down from 10 to 20 feet at a fall. Today the fall cannot be less than 75 feet [i.e., fall in elevation over the distance traversed].

Beaman takes pictures. The boats are let down from rock to rock by lines. Often a boat is held above a chute until a line from it is fixed to some rock below, then the upper line is loosened and away the boat glides or leaps and is swung in below by the lower line.

September 23, Saturday

Let down over falls early. Beaman took picture of "Ross Falls" [?]. After dinner ran narrow rapids tried to take picture of boats running over but failed. Then had a long let down of more than half a mile and camped at an old noon camp of 1869. Take pictures again. Walls of cañon grand beyond description. Towering 2,000 or 2,500 feet. Still the river is filled with huge blocks. A few cottonwood trees at night camp. But the hackberry is the characteristic tree appearing all along just at high-water mark which is from 40 to 50 feet above the present water.

The walls are often nearly vertical on right and grandly craggy on left. River still roars and roars!

September 24, Sunday

Ran round corner on swift water to head of falls and let over the chute very nicely. Then past the old camp [of the 1869 trip] in the amphitheater to the mouth of [a] lateral cañon. Jones, Beaman, and I go up this cañon to some wonderful scenery at its head and return after dark. Gypsum Cañon we have named this. It is formed above by the junction of two others and at its head are grand amphitheaters. Peaks, crags, overhanging cliff, clear pool, and the summit 2,500 feet above.

September 25, Monday

Fred Dellenbaugh and I climb out to Observation Butte [probably North Point of Mancos Mesa]. The day is clear and we see far up the Grand and Green, the Sierra la Sal, the Sierra Abajo, and unknown mountains [the Henry Mountains].[68]

Return by cool spring at cottonwood tree in the gulch just after hard climb over the cliff. In climbing out we attempted to go by the steps of an amphitheater gulch and failed and were compelled to find another way. This we did over some crags and up a crevass. Altitude made today 3,125 feet.

September 26, Tuesday

Ran some wild rapids this morning and let down a good distance. Camped at the bend among grand scenery and took pictures. Run but little more than a mile after dinner, as we stopped for pictures. Camped at mouth of gulch. Thompson and I go up this and select subjects for more pictures.

The cañon walls are about 3,000 feet, nearly vertical, broken by narrow terraces and now and then by side cañons. Up these side cañons the scenery is very fine. Grand amphitheater, etc. At night but a strip of the heavens was seen, say 25° or 30° wide in a crescent shape with five or six constellations Lyra, Casiopia [Cassiopeia], Aquela [Aquila], Delphinus, etc. We sang and told stories until late.

September 27, Wednesday

The scenery still fine as yesterday with the same characteristics. The river fills the channel from wall to wall in many places. Took pictures again today. Passed the terrace where the two sheep were killed two years before and the old night camp, and let down a rapid just before going into camp.

September 28, Thursday

Beaman remains behind to take a picture. I go on to cañon opposite the great bend to the west. This morning the walls are more craggy. We go into camp and Thompson and I try to climb out up a cañon with a beautiful brook and deep clear pools. We fail to reach the summit, cut off by vertical walls of the upper sandstone.[69] Return to camp and start out on the river. Run a very bad rapid at foot of island, the current setting against cliff on the left.

Go into camp just at nightfall at head of a roaring rapid. The cañon here runs to the west. Had a fine view of the rising moon coming up behind the craggs [crags] in the notch of the cañon. A cumulus cloud came before it and reached just above the rocks. A shining silver fringe was on the upper border of the cloud. Behind this the moon rose gradually lighting the walls below us (to the west) and giving a pale weird glow to the waves and breakers and foam and spray of the fall. The cañon walls above were the blackness of gloom but their craggy and serrated outlines set starkly against the lighted sky. From the depths of the gloom faint scintillations of moonlight were reflected from the waters, dimly lighting a few ledges of rock. 'Twas a beautiful, strange, weird view.

September 29, Friday

Up early to see the sun rise through the cañon gap. Thunder shower. This soon brought down a great number of water falls. Some fell a thousand feet.

"The clouds that shattered on you [illegible] narrow walls, etc."

Ran down to Mille Cragg Bend. Thompson and I climb out. Beaman takes pictures.

This morning we saw a rainbow against the walls that was not more than 200 yards away.

Shinumo pottery has been found in many places about the cliffs from the junction [of the Grand and Green Rivers] to this point. Here we find they have lived in a cave with wall in front. Wall laid in mortar. Corn and corn cobs found here at Mille Cragg Bend.

September 30, Saturday

Ran from Mille Cragg Bend to Dirty Devil; so passed through Narrow Cañon, still a great beauty. Passed a number of hot springs. Left Beaman behind to take pictures down cañon and looking out to volcanic mountain.[70]

Arriving at the Dirty Devil.[71] Thompson and I cross, find old Indian trail, follow it along the talus to mouth of what we named Pass [?] Cañon. Then up this cañon 6 miles and returned to camp.

October 1, Sunday

Cache the *Cañonita*,[72] and Jones with me, and Captain Bishop and Steward with Thompson, climb out and get a good view of the Unknown Mountains [Henrys].

Find that we can come around to the north or through the mountains and find a way down to the Colorado River at mouth of Dirty Devil. It would probably be best to come down from foot of the mountain. At night move camp down to mouth of Pass [?] Cañon.

October 2, Monday

Come down to old Shinumo ruins on the left.[73] Stop to explore. Find copious hieroglyphics. Have dinner and run on down to point where we discovered ruins in 1869. Camp there. On digging found that in the angle of the "L" there was an old kiva.[74]

Arrow heads [Arrowheads] and pottery were found at mouth of Dirty Devil at mouth of Pass Cañon and at the two places mentioned above.

October 3, Tuesday

Run down to the camp just below the synclinal axis.[75] Thompson and I climb out at noon. Have fine view of the naked mounds. Find a huge well 40 feet deep with cool clear water at bottom. At certain curves of the river we often obtain a good view of the Unknown Mountains. At noon can see a distant upturned ridge to the west [Water Pocket Fold]. Can trace the synclinal axis with the eye far to the north from point where we climb out at night. Observed some curiously worn valley grooves.

October 4, Wednesday

Our general course this morning is to the south, back and forth across the synclinal axis. The "orange sandstones" are composed of two homogeneous beds separated by stratified beds. The upper homogeneous has few joints and so weathers into mounds. The lower has many and weathers into monuments. The intervening thinly bedded sandstone formed the bottom of the river for several miles this morning, and grooved ledges of the same would spread across the river causing bad rapids and we had much difficulty. The planking of the *Nell* was broken at one ledge and it made a bad leak. Two oars were broken. At noon we stopped some time for repairs. After dinner soon ran out of the hard beds by turning west as the rocks dip east. Ran into the axis of the north and south fold [Waterpocket Fold], crossed it, and along a long straight stretch of river with beautiful view. The left wall 1,400 feet high.

There are many glens along the walls in the thin beds between the two homogeneous [strata]. Glens with springs and oak trees; now and then a cottonwood. There are cañons, shelves, and steps up to them. The river nearly fills the channel from wall to wall. The mounds are higher and no [not] so apparent from the river. Thompson, Steward, and I climb at night into one of these "oak glens" 600 feet above the river.

October 5, Thursday

The Oak Glens still continue as a characteristic of the cañon. The walls are often quite vertical from the water and beautifully rounded at the curves. Sometimes a strip of verdure comes down these [illegible] quite to the water. Pass the San Juan [River] at 11 a.m. Boats go on down to Music Temple while Thompson and I climb up to the west. We obtain a fine view of the mounds of Mt. Seneca Howland [Navajo Mountain] and of Labarynth [Labyrinth] Alcove.[76] Pass up by a cañon to narrow ledge. At noon we go up into the Music Temple. Now we are in Monument Cañon[77] and the Monument Buttes are in view. There are many fine stretches of river with beautiful curves and vertical walls and quiet river. The narrow side cañons are very wonderful and very frequent. At night I go up one of these in a solumn [solemn] way, up through the overarching rocks. The over-arching [overarching] rocks are cut at the circumference of the curves, thus the cañon becomes more curved as it cuts down. I followed up through this cañon until I was cut off by potholes filled with water.[78]

October 6, Friday

Long reaches of quiet river. Finely turned curves at vertical walls are still more noticeable today than yesterday. The monuments too are seen on either hand. This morning we found a little meadow in the cañon, the grass of which had been burned. Tracks of shod horses and of men wearing shoes were seen on the beach. This we suppose to be some of our [illegible] party. At noon have a fine camp on flat rock. Jack catches a large fish and we pull out for the "Crossing of the Fathers."[79] One mile above the crossing found Captain Dodds and two men with our rations. Hamblin had gone to Oraibi. The men's names [are] Ruby and Bonnemort.

October 7, Saturday

Work on map.

October 8, Sunday

Work on map.

[Powell left the party at this point traveling to Kanab, Utah, to set up a base for winter operations and then to go on and bring Mrs. Powell and Mrs. Thompson from Salt Lake City where they had been awaiting the completion of their husbands' trip. The party, under Thompson's charge, proceeded down the river to Lee's Ferry—then called "Lonely Dell"—at the mouth of the Paria River, cached the boats, and moved to Kanab for the winter.

Early the next summer Thompson, Dellenbaugh, and others found their way from Kanab across "Potato Valley" (now Escalante, Utah), discovered the Escalante River, explored the Aquarius Plateau and the Henry Mountains, and found a way to the mouth of the Dirty Devil River. They recovered the cached *Cañonita*, and part of the party took her down the river to Lee's Ferry at the head of Marble Canyon to await the trip through Marble and Grand Canyons. Thompson and the rest of the party returned overland, thence to Lee's Ferry.[80]]

IV. Journal of Trip of 1872 through Marble and Grand Canyons, Lee's Ferry to Kanab Creek, 17 August–8 September

August 17, 1872, Wednesday

Start with the *Cañonita* and *Emma Dean*. At noon take two pictures. Fred climbs out. After dinner make a long portage and find the fragments of the miners' raft. Run down to the next portage and camp 9 ¾ miles from last camp, 11 ½ miles from mouth of the Paria. At this camp the walls of the cañon are 1,000 feet high.

Cañon could be crossed by a suspension bridge. The rapid at camp and the one above are both bad. Fall about 15 feet in each. Let down on the right of both. Here the red beds slope down to the water's edge, covered in places with the loose sands and rocks of the upper beds. In other places the bare red [rock] is seen sloping quite regularly with vertical steps over the harder beds.

August 18, Thursday

I climb to measure the red beds. Return to help with portage. Thompson takes one boat over rocks. I take the other in the water with a line. Have a fine wild whirling ride through a very narrow cañon, plunging down five rapids, until we stop for dinner. Jack and Clem make a picture of the rapids before we start in the morning.

Here at noon just above a bad rapid we are camped on a shelf with overhanging ledge. Three hundred feet of shelving ragged rocks, a slope of 500 feet of the red beds, and 500 feet of smooth vertical wall to summit. Two pictures are made, one looking up and another down. Conclude to run the rapid. All right. Camp just below a bad rapid for night.

August 19, Friday

This day is spent in making portages. At noon obtain some good pictures, two instantaneous ones of a rapid. Two pictures at night just before making portage.

August 20, Saturday

Have two wild rides after making one portage. The walls are of marble below and the entire channel is filled with water. Bad whirlpools. Scenery grand. Stop for pictures at 10 a.m. The foot of the wall comes down to the water's edge with a sharp angle. Camp at night just above Vasey's Fountain.[81] Picture up side cañon.

August 21, Sunday

Scenery very grand today. The marble stands in walls vertical from the water's edge. The forenoon we spend in a vast half-dome-shaped cave 600 feet deep, 900 feet across the mouth, 400 feet high without, and curving down away back.

Take some good pictures at noon a mile below cave. Another picture in the afternoon. Camp No. 92 at Mesquite Bank.[82] The marble below has shown many caves at various heights, colonades [colonnades], and buttresses that set out into the river.

August 22, Monday

The yellowish-green beds are at the base weathering in many beautiful forms. River rapid and full of whirlpools. Ran two great rapids and many smaller ones, let down once. Had several views of the Kaibab. Camp at mouth of Little Colorado.

August 23, Tuesday

In camp I study the shaley rocks. Fred climbs to summit of the marble. Thompson and Jones take observations for latitude. Jack and Clem make pictures.

August 24, Wednesday

Ran down to fault at head of rapid and camped for the day. Thompson and I climb on the south of Kwagunt Valley.[83]

August 25, Thursday

Study rocks of Kwagunt Valley.

August 26, Friday

Study rocks of Kwagunt Valley. Fine climb in afternoon.

August 27, Saturday

Run down to camp on the south side where two sheep were seen. Fine ride. During afternoon Fred and I climb up a trap [rock] cliff and then go up Salt Creek and see the Kwagunt shales again. Return after dark.

August 28, Sunday

Run down a few miles; make one let down about one mile below camp. At ten Thompson and I climb out. Jack and Clem take pictures. Let down after dinner over a long rapid and run down to granite at mouth of Dyke Creek [Cardenas Creek] and let down past bad rapid. Rain.

August 29, Monday

Ran down to the rapid that "must be run," and did run it all right, then down to the "Difficult Portage." Let down around the first corner and camped for night on the rocks. Took pictures before starting over the first rapid for about 50 feet. Grand ride through the granite.

August 30, Tuesday

Let down around the second corner on to a little gravel beach. Then down to the third corner on a rock. Unloaded on the corner and let the boats past. [At] dinner [time] compelled to lay up for repairs. Rain. River still rising. Take the boats up on the rocks—later still higher. Anxious night. River ceases to rise about midnight.[84] Cut mesquite for firewood.

August 31, Wednesday

Launch with great difficulty, the waves beating against the rocks. Run the last part of the rapid all right and stop for repairs on the north side just below. I climb up to sandstones. After dinner run down to mouth of Bright Angel River.[85] The scenery is very grand. River narrow, swift and set into whirlpools by projecting angles of granite, curiously carved, black and somber; and the wonderfully colored and carved rocks above.

Down reaches of the river you can see the summit of the wall a mile high. Pass a curious rapid where the slates are set on edge and the water setting back among the down-turned shelves.

September 1, Thursday

I climb before we start. Take picture at noon from right bank. Run down to portage, make it go into camp and make pictures. Repair boats, etc.

September 2, Friday

Run down to a portage; make it on the left. Run down to another; make it before dinner, except letting down the boats. And then a wild ride through the feldspathic granite to the horn-blendic granite. Make 15 miles.

September 3, Saturday

Two accidents today. In running a rapid in the hornblendic rocks above the Shinumo ruins,[86] the *Emma Dean* was swamped. Letting her float below the great waves she was righted, pulled ashore, and bailed out. Camped for dinner at the Shinumo Ruins above the deep side gulch. Started about 2:30 and had an awe-inspiring ride through two rapids, one setting toward the cliff on the right. The current set the boats against the broken waters along the foot of the cliff with great force so that she seemed to strike against a rock. But passing that in safety we ran around the point to the right in a wild mad current through a narrow gorge that was frightful. On the river sped! bearing us at so swift a rate that no landing could be made; wheeling us round now and then. Trying to stop above a bad rapid, Fred was thrown overboard and failed to get the line ashore. I jumped out on the rock but could not catch her. Then Jones jumped out. On the boat passed. Fred climbed in; she drifted into an eddy behind a huge rock in the middle of the stream and stopped. At last they pulled her ashore where we camped for the night.

September 4. Sunday

Thompson and I spend three hours climbing along the granite under the sandstone to see the river. Find all right for four miles. Away we go with almost railroad speed—a mile in three minutes. Then stop for pictures until after dinner. Then another swift ride, stopping once on the way for a photo, once to examine a rapid and make Camp No. 103.

Sometimes the views are very grand. The water fills the channel from wall to wall in the granite and is very narrow. Sharp angles of granite project into the river and the waves roll up and lash the rocks in deep recesses. So great a volume of water in so very narrow a channel and having a great fall is thrown into mad whirlpools. It boils, foams, lashes the rocks, and rages with great violence. We ran out of the granite at noon today and only little projections into the sandstones are seen from time to time.

September 5. Monday

Made a portage this morning. Then ran a short distance and stopped for photos and had dinner. Then we soon ran into the granite and work commenced. At some places the channel was not more than 25 yards wide. It is difficult to describe the raging of the water through such places.

This afternoon we approached too near a bad rapid and had to pull up again by ropes some 200 yards in order to have room to cross and let down by lines on the opposite side. Camped at night on bed of trap [rock].[87] [Camp] No. 104.

September 6. Tuesday

Spend two or three hours at camp in the study of the rocks. Run down to gulch. I make section up gulch to east. Thompson works on portage and explores a small creek coming in below. Then we run down to Tapeats Creek.[88] The "Arm" [Great Thumb Mesa] is opposite and the [illegible]. Boys take some good pictures.

September 7. Wednesday

Spend forenoon in exploring Tapeats Creek below. Tis a deep gulch in walls of trap [rock], Find Shinumo Ruins. Come down after dinner to cataract. Make picture. Climb over into Surprise Valley [at head of Bonita Creek]. Run down to mouth of Kanab [Creek].

September 8. Thursday

Start up Kanab Creek and explore the upper cañon through to Pink Cliffs. Making sections and collecting fossils. Also study the eruption rocks. Then come down Long Valley [Parunuweap Canyon] and explore the Virgen Cañons to Rockville and return to Kanab by 2 October. Jones, Frank, and George[89] go with us.[90]

Notes

1. Wallace Stegner, *Beyond the Hundredth Meridian: John Wesley Powell and the Second Opening of the West* (Boston, 1962), 123.

2. John Wesley Powell, *Explorations of the Colorado River of the West and Its Tributaries in 1869, 1870, 1871, and 1872, under the Direction of the Secretary of the Smithsonian Institution* (Washington, D. C., 1875).

3. Stegner, *Beyond the Hundredth Meridian;* William Culp Darrah, *Powell of the Colorado* (Princeton, 1951).

4. J. W. Powell, *Report on the Geology of the Eastern Portion of the Uintah Mountains* (Washington, D. C, 1876); G. K. Gilbert, *Report on the Geology of the Henry Mountains* (Washington, D. C, 1877); C. E. Dutton, *Report on the Geology of the High Plateaus of Utah* (Washington, D. C, 1880); Clarence E. Dutton, "Tertiary History of the Grand Cañon District," *United States Geological Survey, Monographs*, No. 2 (1882).

5. Don D. Fowler and Catherine S. Fowler, eds., "The Anthropology of the Nu-ma: John Wesley Powell's Manuscripts on the Numic Peoples of Western North America, 1868–1880," *Smithsonian Contributions to Anthropology*, vol. 14 (1971).

6. John Wesley Powell, *Canyons of the Colorado* (New York, 1895; reprinted as *The Exploration of the Colorado River and Its Canyons*, New York, 1961).

7. Frederick S. Dellenbaugh, *A Canyon Voyage: The Narrative of the Second Powell Expedition Down the Green-Colorado River from Wyoming, and the Explorations on Land, in the Years 1871 and 1872* (New York, 1908; reprinted, New Haven and London, 1962).

8. Herbert E. Gregory, ed., "Almon H. Thompson, Diary," *Utah Historical Quarterly*, vol. 7 (1939), 1–140.

9. William C. Darrah, ed., "Biographical Sketches and Original Documents of the First Powell Expedition," *Utah Historical Quarterly*, vol. 15 (1947), 9–148 (contains newspaper reports written by Powell, the diaries of Powell, George Y. Bradley, and John C. Sumner, and letters to various newspapers written by several members of the party, together with biographical sketches of all the members of the expedition). William C. Darrah, Ralph V. Chamberlin, and Charles Kelly, eds., "Biographical Sketches and Original Documents of the Second Powell Expedition of 1871–72," *Utah Historical Quarterly*, vols. 16–17 (1948–1949), 1–540 (contains the journals of Stephen Vandiver Jones, John F. Steward, and Walter Clement Powell, together with biographical data on these men and the rest of the party). To facilitate cross-reference, hereafter, the writer of the diary and page reference in the appropriate issue of the *Utah Historical Quarterly* is given.

10. J. W. Powell, "Survey of the Colorado River of the West," *House of Representatives, 42nd Congress, 2nd Session, Miscellaneous Document*, No. 173 (1872), 8.

11. Fowler and Fowler, The Anthropology of the Nu-ma.

12. After leaving Powell, Beaman made a trip to the Hopi mesas. He published an account of the river trip and his Hopi visit in 1874: E. O. Beaman, "The Cañon of the Colorado, and the Moquis Pueblos," *Appleton's Journal*, XI (1874), 481–84, 513–16, 545–48, 590–93, 623–26, 641–44, 686–89.

13. The entries for 22–24 May are included as footnotes to Jones' journal, *Utah Historical Quarterly*, vols. 16–17, 24.

14. Jacob Fields was a local resident and outfitter who, together with his family, had entertained Powell and his men prior to the trip.

15. The cabin was owned by "a white man who keeps the ferry at the [Green River] station." Jones, *ibid.*

16. Powell's note.

17. Thompson, Journal, *Utah Historical Quarterly*, vol. 7, 12, records: "Islands in stream all along, . . . When this station was about half run the crew of the *Emma Dean* saw a deer on the bank of a low willow-covered island. The Major shot at but missed it. All the boats landed and we had a good hunt. Got three—two bucks and one doe."

18. In the *Exploration* volume Powell (page 14) wrote: "Let me explain this canyon. Where the river turns to the left above, it takes a course directly into the mountain, penetrating to its very heart, then wheels back upon itself, and runs out into the valley from which it started only a half a mile below the point at which it entered; so the canyon is in the form of an elongated letter U, with the apex in the center of the mountain. We name it Horseshoe Canyon."

19. Named on the first trip: "This is where the river enters the [Uintah] mountain range—the head of the first cañon we are to explore, or, rather, an introductory cañon to a series made by the river through the range. We have named it 'Flaming Gorge' [for the red-colored geological formations]. J. W. Powell, Letter I to *Chicago Tribune*, 2 June 1869, *Utah Historical Quarterly*, vol. 15, 75–78.

20. Named on the 1869 trip, "for a kingfisher we saw at mouth of Fork." Bradley, Journal, *Utah Historical Quarterly*, vol. 15, 33.

21. It is not clear what Powell is referring to here. He may mean the "Bad lands" area north of Henry's Fork.

22. Powell's name for a peak on the Wyoming-Utah border; its present name is uncertain since the location is so vague.

23. Named in 1869: "The bluff opposite our camp we call 'Beehive Point' from its resemblance in shape to a straw beehive." Bradley, *ibid.*

24. Named in 1869: "Named the cañon 'Red Cañon,' for it is chiefly red sandstone." Bradley, *ibid.* The upper section of Red Canyon is now called Hideout Canyon.

25. These are negative numbers assigned by Powell to the photographs.

26. Thompson, *Utah Historical Quarterly*, 15, writes: "There lay on

the slope, half buried in the sand, a boat, which once belonged to a party of prospectors who two years ago tried to descend from Green River City to Brown's Hole in boats. They got as far as just above our camp when one of their number was drowned. The party disbanded, abandoning their boats and made the rest of their way overland to their place of destination."

27. *Ibid.*, 16: "When we got back the Major read 'The Lay of the Last Minstrel' aloud. It has been a most enjoyable Sunday."

28. Now called Trail Creek and Allen Creek, respectively.

29. In 1869 Powell reported: "On a rock, by which our trail [for the portage] ran, was written "Ashley," with a date, one figure of which was obscure—some thinking it was 1825, others 1855. I had been told by old mountaineers of a party of men starting down the river, and Ashley was mentioned as one; and the story runs that the boat was swamped and some of the party drowned in the cañon below. This word 'Ashley' is a warning to us and we resolve on great caution. 'Ashley Falls' is the name we have given the cataract." General William Henry Ashley (the rank derived from a commission in the Missouri state militia) was a cofounder of the Rocky Mountain Fur Company. In 1825 he explored the Wyoming and Utah areas with his trappers. On 22 April of that year he and six men began a float trip down the Green River from the mouth of Sandy Creek in a buffalo-hide bull boat (later a dugout canoe). They arrived at the mouth of the Duchesne River in late May, a truly remarkable feat given the boats used and the fact that Ashley could not swim. Ashley's diary of the trip is contained in Dale L. Morgan, ed., *The West of William H. Ashley, 1822–1838* (Denver, 1964), 104–15.

30. Lena was Bishop's sister. Bishop, *Utah Historical Quarterly*, vol. 15, 170.

31. These were sent out with Messrs. Bacon and Harrell, cattlemen who were camped at Brown's Hole, and who had brought mail for the party. Thompson, *Utah Historical Quarterly*, 18; Bradley, *Utah Historical Quarterly*, 170.

32. There is no entry for 10 June. Powell does not mention that Frank Richardson left the party at this point. Bishop, *Utah Historical Quarterly*, 170, wrote: "Frank will leave us here and go back to Green River Station, as he was found in no way suitable for the trip. It seems rather sad to part with him after all. His heart is kind and although he is not of much value, yet he is one of us."

33. Named for the large flocks of swallows living there. W. C. Powell, *Utah Historical Quarterly*, vols. 16–17, 271.

34. The party was still passing through Brown's Park, which extends along the river for some miles.

35. Named on the 1869 trip. Darrah, in Bradley, *Utah Historical Quarterly*, 40, Note 13, says the name was "Suggested by Southey's poem, which the Major knew by heart." Robert Southey's (1774–1843) poem "The Cataract of Lodore," from *The*

Complete Poetical Works (New York, 1851), 175, *is* apropos:

> *Till in this rapid pace,*
> *On which it is bent,*
> *It reaches the place,*
> *Of its steep descent,*
> *The Cataract strong*
> *Then plunges along,*
> *Striking and raging,*
> *As if a war waging,*
> *Its caverns and rocks among . . .*

36. Jones, *Utah Historical Quarterly*, 37, calls this Turtle Back Mountain; Thompson, *Utah Historical Quarterly*, 19, refers to it as Turtle Creek Mountain.

37. Mail for the party was brought overland by Harrell's men. Thompson, *ibid.*

38. The grotto was named for J. F. Steward's daughter, according to W. C. Powell, *Utah Historical Quarterly*, 273, and Dellenbaugh, *A Canyon Voyage*, 35; Steward, *Utah Historical Quarterly*, vols. 16–17, 193, simply wrote: "During the afternoon we spent a little time in Winnie's Grotto, and the topographer [*sic*] took several views thereof."

39. During the 1869 trip the *No Name* was lost in these rapids and its crew nearly drowned. Bradley, *Utah Historical Quarterly*, 36, described the accident as follows: "June 8, 1869 . . . as we advanced [the rapids] grew worse until we came to the wildest rapid yet seen. I succeeded in making a landing in an eddy just above where the dangerous part began. So did one other of the heavy boats, but one (the "No Name") with three men in it with one-third of our provisions, half of our mess kit, and all three of the barometers went over the rapid and though the men escaped with their lives yet they lost all their clothing, bedding, and everything except shirt and drawers, the uniform in which we pass all rapids."

40. A camp ground of the first trip. Several of these are noted throughout the text.

41. An island in the river on which a fragment of the *No Name* had lodged and from which the barometers were recovered. Bradley, *ibid.* During the second trip, other tools and fragments were found below the island. W. C. Powell, *Utah Historical Quarterly*, 274–75.

42. Named for the constellation *Lyra*, the brightest star of which is *Vega*.

43. Named for William H. Dunn, who, with the Howland brothers, left the first party in the Grand Canyon. They were killed by Indians.

44. Bishop, *Utah Historical Quarterly*, 112, also noted the monument. The editor of Bishop's journal indicates that the monument has never been rediscovered. Bishop, *ibid.*, Note 10.

45. Dellenbaugh, *A Canyon Voyage*, 44, wrote: "It took us, with hard work, till two o'clock to get past Triplet Falls by means of a double portage. About a half mile below this we were confronted by one of the worst looking places we had yet seen, and at the suggestion of Steward it received the significant name of 'Hell's Half Mile.' The entire river for more than a half a mile was one sheet of white foam. There was not a quiet spot in the whole distance and the water plunged and pounded in its fierce descent and sent up a deafening roar."

46. The park [Pat's Hole] is at the confluence of the Green and Yampa Rivers. The name recognizes the acoustical properties of Echo Cliff [Steamboat Rock] on the west bank of the green. Dellenbaugh, *ibid.*, 49, reports "a distinct echo of ten words." Other estimates are smaller.

47. Jones, *Utah Historical Quarterly*, 43, "made camp on right bank in a beautiful park. From the appearance this was a resort of the bear so we named it 'Grizzly Park,' and a beautiful spring near the river the same."

48. Powell later changed the name to "Bishop Creek" in honor of F. M. Bishop. It is now known as "Jones Hole Creek," for S. V. Jones.

49. Thompson, *Utah Historical Quarterly*, 78–87.

50. The Book Cliffs form the lower, and the Brown or Roan Cliffs the upper escarpments of the Tavaputs Plateau which is bisected by the Green River.

51. Jones, *Utah Historical Quarterly*, 76, wrote: "Beaman went up the river and took some views of a lateral cañon on the other side of the stream. Because it is divided into three parts he called the pictures 'Trin Alcove from across the river.'"

52. This refers to a butte near the mouth of the San Rafael River. Dellenbaugh, *A Canyon Voyage*, 102, writes: "Many of these buttes were beautiful in their castellated form as well as because of a picturesque banded character, and opposite our dinner camp . . . was one surprisingly symmetrical, resembling an artificial structure. I thought it looked like an art gallery, and the Major said it ought to be named after the artist, so he called it 'Dellenbaugh's Butte,' then and there."

53. This and the following paragraph are reprinted as a footnote in Jones's journal, *Utah Historical Quarterly*, 47.

54. Powell's note.

55. A deep meander in the river. Thompson, *Utah Historical Quarterly*, 45, notes: "Ran around the south bow of the 'Knot.' Ran 5 ⅛ miles to get 1000 feet."

56. An abandoned U-shaped river meander.

57. The Orange Cliffs are part of an escarpment bisected by the Green River near its confluence with the Grand River.

58. The "Needles Country," now part of Canyonlands National Park.

59. Thompson, *Utah Historical Quarterly*, 41, calls them "a Leguminoise."

60. Jones, *Utah Historical Quarterly*, 81: "On the steep walls were incrustations of salt that evidently must have percolated through the rocks from the Triassic [formation] above."

61. In Volume 6 of Powell's journals (back of page 148) he records that, "Shi-no-mo is the name of the people of all these towns" (that is, the villages on the Hopi mesas in northern Arizona), "Moqui is the name given to the three last" (Ho-pi-ki, Shi-choam-u-vi, Ha-no-ki). "Moqui" or Moki is apparently a hispanization and, later, anglicization of the Ute-Southern Paiute term "moq́-wi'tc." Edward Sapir, "Southern Piute Dictionary," *Proceedings of the American Academy of Arts and Sciences*, LXV, 537–770, 1931. "Moqui" was commonly used in the Colorado Plateau area to refer to the Hopi and to Puebloan archeological sites. It is still used in the latter sense by local residents. Here, and below, Powell used the terms "Shinomo" and "Moqui" to refer to Puebloan archeological sites. The site here noted was probably built by carriers of the Fremont cultural tradition who occupied the area north of the Colorado River from about AD 800 until perhaps 1300. C. Melvin Aikens, "Fremont-Promontory-Plains Relationships," *University of Utah Anthropological Papers*, No. 81, (1966).

62. The "climb out" was made by part of the party in the morning; the "run down" to the confluence of the Grand and Green Rivers to join the rest of the party was made later in the day. Thompson, *Utah Historical Quarterly*, 49.

63. Thompson, *ibid.*, records: "The party that stayed behind climbed out at the 'Moquis trail.' . . . They found a good many broken arrowheads and the chips from their making, pieces of pottery, and under a shallow rock, carefully covered by a thin, flat rock, Andy found a whole jar. It will hold about four gallons and was partially filled with split willows, such as are used in making trays. The willow splints were tied up by hemp from the nettle or bark." Jones, *Utah Historical Quarterly*, 82, adds: "The Major recognized the kettle [jar] as the work of the Chi-ne-mos and the willows as the material from which they make Pe-too or Bread trays."

64. At a point about one mile up the Green from the confluence. Jones, *ibid.*, 83.

65. Jones, *ibid.*, 84, writes: "Major and I . . . struck down the Colorado. Went for three miles, then turned to the northwest and walked for hours through lovely valleys and alcoves surrounded by steep walls of white rock. At length, after climbing through a steep narrow gorge we came in sight of a splendid view. Towers, columns, pinnacles, turrets, domes in every direction. Wandered until the sun warned us of approaching night. The Major seemed like one entranced. Said we would call this the 'Sin-av-ton-weap.'"

In Ute and Southern Paiute mythology, with which Powell was familiar, "Sinav" or "Shinauav" is a creator figure. "Tuweap" means land, or country, or region. Hence: "Creator's country," or, "A God's Country." The northern part of this area is now called "The Maze," the southern section is part of the area termed "Ernie's Country."

66. They were probably in Len's Canyon, a tributary of Lower Red Lake Canyon that enters the Colorado at the place specified.

67. This paragraph appears as a footnote in Jones's journal, *Utah Historical Quarterly*, 85, Note 57.

68. Later named by Powell for Joseph Henry, first Secretary of the Smithsonian Institution.

69. Thompson, *Utah Historical Quarterly*, 52, records that they called the creek "Failure Creek, in remembrance of our failure."

70. Probably referring here to one of the Henry Mountains.

71. Dellenbaugh, *A Canyon Voyage*, 133–34, wrote: "Landing on its west bank, we instantly agreed with Jack Sumner when on the first trip he had proclaimed it a 'Dirty Devil.' Muddy, alkaline, undrinkable, it slipped along between the low walls of smooth sandstone to add its volume to that of the Colorado."

72. Since Powell had been unable to get supplies to the mouth of the Dirty Devil, and his stocks were running low, he had previously decided to cache one boat here and move quickly through Glen Canyon to the Crossing of the Fathers. The following summer Thompson, having found an overland route to the mouth of the Dirty Devil, recovered the boat and it was taken to the mouth of the Paria River, thence through the Grand Canyon later that summer (see below).

73. This archeological site was located just downstream from the settlement at the mouth of White Canyon prior to its inundation by Lake Powell in 1963. It is recorded by the University of Utah. Ted D. Weller, San Juan Triangle Survey, in Don D. Fowler et al., "The Glen Canyon Archeological Survey," *University of Utah Anthropological Papers*, No. 39, Part 2 (1959), 543–669.

74. This archeological site, named Loper Ruin after Bert Loper, a homesteader who built a cabin nearby in the early 1900s, was excavated by the University of Utah in 1958 and 1959. The excavations verified Powell's assertion that a depression in the "angle of the L," i.e., between the two wings of the structure, was probably a kiva (an underground ceremonial chamber built by Puebloan peoples), though an unfinished one. William D. Lipe, "1958 Excavations, Glen Canyon Area," *University of Utah Anthropological Papers*, No. 44 (1960), 129–32.

75. A point at which the Moenkopi and Chinle geological strata in the canyon walls dip below river level.

76. Seneca Howland was one of the men who left the first river party at Separation Rapids and was killed by Indians. The name Navajo Mountain has been retained. "Labyrinth Alcove" may refer either to Nasja Creek or Mystery Canyon.

77. In an editorial footnote to Thompson's journal, *Utah Historical Quarterly*, 54, Note 30, Gregory says: "The stretch of river from the mouth of the Fremont [Dirty Devil] to the mouth of Trachyte Creek was first called Mound Canyon bv Powell, and its extension to the mouth of the Paria, Monument Canyon. On published maps the two are combined as Glen Canyon." The canyon is now inundated by Lake Powell, formed behind the Glen Canyon dam.

78. This is probably the upper channel of Music Temple which was accessible over the rounded cliff upstream from its mouth; if, however, Powell was across the river he might mean Hidden Passage. Both fit the description.

79. Named for the ford across the river used by Father Escalante's party during its overland explorations of what is now Utah and Arizona in the 1770s. Also called "Ute Ford." See Herbert E. Bolton, ed., "Pageant in the Wilderness, the Story of the Escalante Expedition to the Interior Basin, 1776," *Utah Historical Quarterly*, vol. 18, (1950).

80. Thompson, *Utah Historical Quarterly*, 78–87.

81. Named "Vasey's Paradise" in 1869 by Powell for George W. Vasey (1822–1893), a botanist from Illinois Normal University who accompanied Powell on his 1868 trip to the Rocky Mountains. See *The Professor Goes West: Illinois Wesleyan University—Reports of Major John Wesley Powell's Explorations: 1867–1874*, comp. and ed. by Elmo Scott Watson (Bloomington, Illinois, 1954), 8. Vasey later became a botanist for the United States Department of Agriculture.

82. A sand bar with a growth of mesquite bushes on it.

83. There is some confusion here. Gregory, in Jones, *Utah Historical Quarterly*, 148, Note 130, indicates the position of Kwagunt Creek as above the mouth of the Little Colorado River, and that of Chuar Creek at the place here indicated.

84. Jones, *ibid.*, 150, reported that the river rose 6 feet.

85. In his 1875 report, Powell, *Explorations*, 86, writes, under the date 16 August 1869: "We have named one stream, away above, in honor of the great chief of the "Bad Angels" [Dirty Devil River], and, as this is in beautiful contrast to that, we conclude to name it "Bright Angel." Yet in Powell's 1869 journal, *Utah Historical Quarterly*, vol. 15, 129, he calls it "Silver Creek." William Culp Darrah, in a footnote to Bradley's journal, 65, Note 43, writes: "The name 'Bright Angel' was not bestowed upon this stream until December 1869 when the Major used it on the lecture platform to give romantic contrast to 'Dirty Devil.'"

86. These are apparently the same ruins Powell recorded on the first trip: "Aug. 20 [1869]...Came at night to the 'old red' [sandstone]. Found the remains of old Moqui village on bank; stone houses and pottery.

    (Found some remains at Silver Creek [Bright Angel Creek])." Thompson's diary, *Utah Historical Quarterly*, 96–97, indicates that the ruins are about 23 miles downstream from Bright Angel Creek. The site is at the lower end of Stanton's Switchyard and was rerecorded in 1962 by Robert C. Euler (personal communication).

87. A dark-colored igneous rock.

88. Named for a Southern Paiute Indian whom Powell had met earlier in the year.

89. Frank and George were Southern Paiute Indians; the former was a leader of the Kaibab band and was better known as Chuarumpeak.

90. The editors are indebted to John C. Ewers for comments on the manuscript, to Margaret Blaker for archival assistance, and to Robert C. Euler for information on archeological sites in the Grand Canyon. Work on this and other Powell manuscripts was made possible by a National Research Council postdoctoral fellowship at the Smithsonian Institution awarded to the senior editor, and a grant to both editors by the National Endowment for the Humanities. This support is acknowledged with gratitude.

PART 2

"Photographed All the Best Scenery"

Jack Hillers's Diary of the Powell Expeditions, 1871–1875

John K. Hillers

INTRODUCTION BY DON D. FOWLER

In May 1871, John K. (Jack) Hillers chanced to meet Major John Wesley Powell in Salt Lake City. Powell and his brother-in-law Almon Harris Thompson were in the city to get their wives settled and to make final preparations for a second boat trip down the Green and Colorado Rivers. One of the men scheduled to accompany the expedition could not go and Powell was looking for a replacement. Somehow Powell met Hillers, who was working in the city as a teamster. Powell liked the looks of the tall, red-haired Hillers and offered him the boatman's job. Hillers accepted.

The meeting marked the beginning of a lifetime association between Powell, Thompson, and Hillers. In subsequent years Powell was to become the director of both the Bureau of American Ethnology and the United States Geological Survey; Thompson was to become the chief topographer of the Geological Survey; Hillers was to become the chief photographer for Powell's Geological and Geographical Survey of the Rocky Mountain Region, and, in 1881, the chief photographer for the United States Geological Survey. Through the years he made many series of now classic photographs of the Indians and the geological formations of the Colorado Plateau and the American Southwest. His work ranks with that of other great nineteenth century photographers of the American West, including Timothy O'Sullivan, William H. Jackson, Charles R. Savage, and Christian Barthelmess.[1]

Hillers kept a diary of the events of the trip down the Green and Colorado Rivers in 1871 and 1872, of some of his work for the Powell Survey in 1874, and of a trip to Indian Territory in 1875. The present volume presents Hillers's diary, together with a selection of his photographs made between 1872 and 1879. The volume is thus a contribution to the history of the Powell expeditions and the American West and to the history of American photography.

The Powell Survey

Beginning with the Lewis and Clark expedition in 1803–05, the United States government sponsored many expeditions to the American West to make maps and to collect data on resources and Indian tribes. Prior to the Civil War these expeditions were conducted by the United States Army Corps of Topographical Engineers.[2] After the Civil War, federal support for western exploration continued, but a new type of survey developed. With the exception of the Geographical Survey West of the One Hundredth Meridian, led by Captain George M. Wheeler of the United States Army, the new surveys were under civilian control.

Ultimately there were four federally sponsored surveys, including the Wheeler Survey, operating in the West. The three civilian surveys were the Geological and Geographical Survey of the Territories, led by Ferdinand Vandeveer Hayden; the Geological Exploration of the Fortieth Parallel, led by Clarence King; and the Powell Survey, which had several names but was best known as the Geological and Geographical Survey of the Rocky Mountain Region.[3] The Powell Survey was the smallest and the last on the scene, but it, and its director, were to have enormous impact on federally sponsored scientific research, land use practices, and conservation practices. In 1879 all four surveys were merged into the United States Geological Survey.

The story of Powell's survey and subsequent career as director of two major government scientific agencies, the Bureau of American Ethnology and the United States Geological Survey, is well known.[4] Here we will only sketch Powell's career in its relation to Hillers and the work Hillers performed under Powell's direction.

Powell emerged from the Civil War with one arm and the rank of major to become a professor at Illinois Wesleyan University. In 1867 he led a party of students and friends on an expedition to the Rocky Mountains.[5] In 1868 he led a second party to the same area, but in the fall he, with his wife and a few others, remained behind to explore the upper reaches of the Green River in western Colorado, eastern Utah, and southern Wyoming. During the winter Powell formulated plans to explore the Green and Colorado Rivers in boats.[6] In 1868–69 only the Green River above the Uintah Basin and the lower Colorado below Black Canyon (the present site of Hoover Dam) were known and mapped. The area between, comprising most of the deeply dissected Colorado Plateau, was virtually unknown. Some brief attempts had been made to run the upper Green River as early as 1825, and rumors abounded of giant whirlpools and rivers disappearing underground—but there were few facts.[7]

Powell resolved to explore the river system. He gained support from several institutions in Illinois, and on May 24, 1869, he and nine men in four boats set out from Green River Station, Wyoming Territory. On August 30, 1869, Powell and six men in two boats emerged from the foot of the Grand Canyon at the small Mormon settlement of Callville, Nevada. Along the way two boats were lost or abandoned and three men had left the party in the Grand Canyon. These men were later killed by Shivwits Indians.

The river trip made Powell a national hero—and gained him a congressional appropriation to continue his explorations. Much of the data collected during the first trip was lost and a shortage of supplies had forced the party to hurry. But Powell had learned that the rivers were navigable by properly constructed small boats. As difficult and fearful as the rapids and falls were in many places, Powell and his men had learned how to run them or portage around them.

Powell now planned a second trip, this time with supplies cached along the way by overland pack teams and with adequate scientific apparatus and time to use it. The 1869 trip had been an adventure into the unknown; the 1871 trip was a scientific expedition designed to collect new geological, hypsometric, and other data—and to photograph for the first time the fantastic Canyon Country.

In May 1871, Powell and Almon Harris Thompson, Powell's brother-in-law and chief topographer, arrived in Salt Lake City, met Hillers, and hired him.

Powell, Thompson, and Hillers arrived at Green River Station on May 16, 1871, to find the other members of the river party waiting for them. Hillers's diary begins on that date.

Following the second river trip Powell was able to gain continued congressional appropriations for his survey. He, Thompson, and others spent most of the 1870s studying and mapping the geology of Utah and Arizona and producing reports which remain classics in the field. Hillers accompanied the various field parties as photographer. Many of his photographs were used as the bases for the engraved illustrations by Thomas Moran and others in the geology reports produced by members of the Survey.[8]

Powell's interests included anthropology as well as geology. He had begun studying the Ute and Southern Paiute Indians of the Colorado Plateau in 1868 and became competent in their languages. Whenever he got the chance he recorded ethnographic and linguistic data.[9]

In May 1873, Powell was appointed a Special Commissioner of Indian Affairs to look into a number of Indian problems in Utah, Idaho, northern Arizona, and Nevada. Together with George W. Ingalls, Powell met various Indian delegations in Salt Lake City and then traveled south meeting various Indian bands and delegations along the way.[10] Hillers met Powell in Kanab, Utah, and accompanied him to St. George, Utah, and Moapa and Las Vegas, Nevada. At these points, Hillers made a series of photographs of various Southern Paiute Indians. These photographs provide valuable ethnographic data on the Indians.[11] But some few of them are not true ethnographic records. Powell had collected some buckskin clothes from the Northern Utes of the Uintah Basin in 1868. These he brought with him to southern Utah and dressed some of the Southern Paiutes in them. In the photographs, one, and possibly two, Southern Paiute women are dressed in Northern Ute beaded buckskin dresses. In one photograph, the word "Colorado" and a museum accession number are clearly visible on the bodice of the dress. In some of the photographs the men are wearing feather headdresses. Such ornamentation was not indigenous to the Southern Paiute, who usually wore close-fitting caps. There is some evidence that the headdresses were made for the occasion under the direction of Ellen Thompson, Pow-

ell's sister. Some of the poses in the photographs are highly stylized nineteenth century "art" poses. There is also an element of cheesecake in some of the photographs: some Indian women were posed with one breast just visible.

Although Powell recognized the ethnographic value of the photographs, he had other uses for them in mind as well, seeing them primarily as a source of income. There was a substantial market in the 1870s for stereographs for the stereoscopes found in most nineteenth century homes.[12] Powell and Beaman had made an arrangement to split the proceeds from the stereographs produced during the river trip. But when Powell and Beaman disagreed and Beaman left in January 1872, Powell bought out Beaman's interest. After Hillers became Survey photographer, he, Powell, and Thompson entered into an agreement to share the proceeds of the sale of stereographs. Powell received forty percent, Thompson thirty percent, and Hillers thirty percent. Powell took Hillers's share from the proceeds of Beaman and Fennemore negatives. There are few figures on the proceeds from these sales, but it is known that they totaled $4,100 for the first six months of 1874. Darrah notes that a standing joke in the United States Geological Survey in the late 1880s was that Powell had paid off the mortgage on his house in Washington through sales of the views.[13]

After 1873 Powell established a regular office in Washington for his Survey. Hillers and other Survey employees thereupon spent part of the year in Washington. Sometimes Hillers accompanied Powell on lecture tours.[14]

In 1875 Powell was asked to direct the collection of Indian artifacts, information, and photographs for the Smithsonian Institution exhibit at the 1876 Philadelphia Centennial Exposition. As a part of this effort, Powell sent Hillers to Oklahoma, then Indian Territory, to take pictures of various Indian tribesmen. The last portion of Hillers's diary is devoted to this trip.

In 1879 Powell succeeded in getting the Bureau of Ethnology (after 1894, the Bureau of American Ethnology) established as a part of the Smithsonian Institution, with himself as director.[15] Hillers was hired as Bureau photographer. One of Powell's first acts was to send Hillers, together with

James Stevenson and Frank Hamilton Cushing, to Arizona and New Mexico to survey archeological ruins and photograph the Pueblo Indians.[16] Several of Hillers's best pictures derive from this trip.

Powell had had a behind-the-scenes hand in the formation of the United States Geological Survey, also formed in 1879. In 1881, the director, Clarence King, resigned and Powell was appointed to replace him, an appointment he took on in addition to his directorship of the Bureau of Ethnology. Hillers was transferred to the Survey payroll (which always had more money than the Bureau) and became the chief photographer of the Survey at a salary of $1,800 per year.[17] He remained in that capacity until his retirement in 1900. Between 1881 and 1894, when political pressures forced him to resign as Director of the Survey, Powell ran both the Survey and the Bureau from the same office. Hence, Hillers was usually available to take portraits of visiting Indian delegations for the Bureau.

Hillers's Life

There is little biographical data about Hillers's early life.[18] He was born in Hanover, Germany, in 1843 and came to the United States at the age of nine. At the beginning of the Civil War he enlisted in the New York Naval Brigade, but later transferred to the army. He saw action at Petersburg, Cold Harbor, and Richmond. At the end of the war he re-enlisted and served in various Western garrisons.

In 1870 Hillers resigned from the army to accompany his brother Richard to San Francisco. Returning eastward in 1871, he stopped in Salt Lake City and went to work as a teamster. There, as we have seen, he chanced to meet Powell and Thompson and signed on their river expedition. Hillers was well liked by all. His capabilities, sense of humor, and evenness of temper earned him the nickname of "Jolly Jack," though he was most often called "Bismark" [sic] for his German background. Hillers was especially close to Powell. As Darrah has written, "Hillers' buoyant spirit and, at times, ribald sense of humor, endeared him to the Major. He was the one man who could be breezy and flippant with his chief. Hillers alone dared write to the Major, 'Love to Mrs. Powell and kisses for the girls....'"[19]

Hillers's transition from boatman and handyman to chief photographer of the Powell expedition took place during the 1871 river trip and in the following year. Powell had hired E. O. Beaman, a professional photographer from New York City, to make photographs during the river trip. Beaman's equipment—large format camera, portable darkroom, chemicals, and large glass negatives for the collodion wet-plate process then used—weighed nearly a ton. Powell appointed his cousin, Walter Clement (Clem) Powell, Beaman's assistant.

During the river trip Hillers became interested in photography. Frederick S. Dellenbaugh, a member of the expedition, in a letter to Robert Taft, the historian of early American photography, explained how Hillers became a photographer:

> Jack Hillers was engaged by Powell for the second trip, in Salt Lake City, to pull an oar and to help generally. He knew nothing then about photography, but he became much interested as we went on down Green River.
>
> When we had arrived at the mouth of the Green...Hillers asked me about photography—about the chemical side. I explained about the action of light on a glass plate coated with collodion and sensitized with nitrate of silver, the bath to eliminate the silver in hypo, and so on.
>
> "Why couldn't I do it?" he said. I replied that he certainly could, for he was a careful, cleanly man, and those were the chief qualities needed. I advised him to offer his help to Beaman whenever possible...and perhaps Beaman would let him try a negative. He did, and in two or three weeks had made such progress that he overshadowed Clement Powell.
>
> When Beaman left at the end of our first season of river work, the photography fell on Clement. He made a trip with Hillers as assistant and returned with nothing. He declared Beaman had put hypo in the developer. Beaman...said he had not done so—that the trouble was Clem's carelessness.

Meanwhile, Powell had sent to Salt Lake City to see if a photographer could be found there who would come down [to Kanab, Utah,] and who would go through the Grand Canyon with us the next summer. James Fennemore came. He was an excellent photographer and a genial fellow. . . . He was good to Hillers and gave him much instruction, with the result that Hillers became expert in the work, became assistant, in fact. . . .

Hillers did some excellent work with Fennemore's guidance. When we were preparing to enter the Grand Canyon in the summer of 1872 Fennemore was taken sick . . . [and] could not proceed. And we could not find anyone else who desired to do the Grand Canyon with us. This left us with only seven men. One boat had to be left behind. . . . Well what I am coming to is this; there was nothing for it but to make Jack Hillers photographer-in-chief. He was equal to the job. In spite of enormous difficulties, great fatigue, shortage of grub, etc., he made a number of first class negatives.[20]

Hillers made several field trips in the 1880s. While in the field he was, as always, resourceful. Captain John Gregory Bourke, the famed ethnologist and aide-de-camp to General George Crook, ran into Hillers and a Survey crew in northern Arizona near the Mormon settlement of Sunset in the summer of 1881. Libations were clearly in order. As Bourke reports, "Hillers reappeared with a mixture of ginger and whiskey. The ginger was all right, but the whiskey, from the Mormon town of Brigham, was as vile as Arizona could produce."[21]

After 1885 increasing administrative duties kept Hillers in Washington most of the time. In 1886 he received a raise in pay—to two thousand dollars per year, a sum he continued to earn until his retirement. On May 1, 1900, Hillers retired from the Survey, but continued to work on a per diem basis until 1919.[22] In September 1902, Hillers had the sad duty of serving as a pallbearer at Powell's funeral.[23] The close relationship between the two men had continued until the end.

John K. Hillers died in Washington, D.C., in 1925 and was buried in Arlington National Cemetery, near the graves of Powell and Thompson.

Hillers's Photographs

Jack Hillers made several thousand negatives of anthropological and geological subjects in his twenty-nine years with the Powell Survey, the Bureau of Ethnology, and the Geological Survey. Many of these are deposited in the National Archives and the Smithsonian National Anthropological Archives in Washington, D.C. The Denver Public Library also has a large collection of Hillers's photographs and stereographs.

Wallace Stegner thought Hillers was the first to photograph the Grand Canyon, but the honor goes to Timothy O'Sullivan with the Wheeler Survey in 1871. Hillers's first Grand Canyon photographs were made the following year.[24] His ethnographic series of photographs of Southern Paiute Indians, although some of his models are posed in artificial attitudes, are a valuable source of information, as are his photographs of Zuni and various Oklahoma Indians.

Hillers pioneered the making of large photographic transparencies on sheets of glass up to 4 × 5 feet in size. Some of these were hand colored. They were used in several of the Smithsonian Institution and Bureau of Ethnology exhibits at national and international fairs and expositions.

Hillers was first and foremost a craftsman. His photographs are technically excellent, a remarkable achievement given the wet-plate negative process in use until the 1880s. But in many of his photographs he went beyond technical excellence. In composition and content many of his photographs are works of art.

Hillers's Diary

Most of the known diaries and journals of the Powell river expeditions of 1869 and 1871–72 have previously been published.[25] Hillers's diary was deposited in the Bureau of American Ethnology archives (now incorporated into the Smithsonian National Anthropological Archives) in 1952 by Mrs. J. K. Hillers, Jr., Hillers's daughter-in-law. It consists of two small leather-bound books and a number of loose sheets. In 1952 Matthew W. Stirling, then Chief of the Bureau of American Ethnology, collated and organized the loose sheets and had a typescript copy made. In 1968 the present editor checked the typescript against the original diaries. The script is small and sometimes difficult to read, some words being illegible.

Hillers apparently started his diary some weeks after the beginning of the river trip. Stephen Vandiver Jones's diary for August 9, 1871, contains the following passage: "Hillers has procured a book and spent the time today in writing up from the start. Believe each member of the party is keeping a journal with the exception of Hattan."[26]

On August 9, 1871, Hillers had just returned from the Uintah Indian Agency with supplies for the river party. He may have acquired the book at the agency. After the river party reached Kanab, Hillers apparently mailed sections of the diary to his brother Richard in San Francisco, and later sent other sections to him. An undated page in the materials contains the note, "Dick, don't lose any leaves out of this, be careful. You can amuse yourself reading it then put it in my trunk."

The diary covers the period from May 16, 1871, through October 26, 1872 (with some breaks), as well as five days (September 11–15) probably in 1873, and the period May 1 through June 10, 1875. Included with the 1875 material is a portion of an undated letter to Hillers's brother describing a meeting with a charming Indian girl in Indian Territory. Unfortunately, the last section of the letter is missing and we do not learn of the outcome of the meeting.

In editing the diary for publication care has been taken to retain Hillers's original spelling. Editorial inserts in brackets correct or clarify proper names as necessary.

Acknowledgments

At the time the Powell expedition journals were being published in 1947–49, Hillers's diary remained in his family's possession. Later, the late Mrs. John K. Hillers, Jr., donated the diary to the Smithsonian Institution. In 1968 she kindly granted permission to me to publish the diary. Special thanks are due Mrs. Hillers for her kind permission and for permitting my wife and me to examine the Hillers family scrapbooks and memorabilia.

Between 1967 and 1973, Catherine S. Fowler and I edited for publication a series of manuscripts, including the Hillers diary, relating to John Wesley Powell, the Powell expeditions, and the Bureau of American Ethnology. The manuscripts are on deposit in the Smithsonian National Anthropological Archives, Washington, D.C. The work was made possible by a grant from the National Endowment for the Humanities and a National Academy of Sciences Post-doctoral Research Fellowship at the Smithsonian in 1967–68. This support is acknowledged with gratitude. The late John C. Ewers, the late Margaret Blaker, and the late Clifford Evans, and Joanna Scherer of the Smithsonian were most helpful with the Hillers diary. The late Nell Carico of the United States Geological Survey, Marilyn Seifert of the Latter-day Saints Church Historian's Office, Everett L. Cooley of the University of Utah, Jack Haley of the University of Oklahoma, Rella Looney of the Oklahoma Historical Society, the late Vilate Hardy Gubler of La Verkin, Utah, and the late Juanita Brooks of St. George, Utah, all supplied information to aid in annotating and verifying details of the diary. My thanks to each of them. Finally, a special thanks to my wife, Catherine S. Fowler, for editorial, critical, and general support.

JACK HILLERS'S DIARY OF THE POWELL EXPEDITIONS, 1871–1875

1871: John K. Hillers U.S. Survey of Colorado River 1871

Green River, May 16, 1871

Left Salt Lake with Major Powell and Prof. Thompson at 5 a.m. to accompany them on the Exploring Expedition of the Green and Colorado Rivers. Got here at 6 a.m. where I found the rest of our party. The whole party stands as follows: Major [John Wesley] Powell, Geologist; Prof. [Almon Harris] Thompson, Astronomer and Topographer; [Francis Marion] Bishop (Cap) and [Stephen Vandiver] Jones (Deacon), Assistents [sic]; [John F.] Steward, Assistent Geologist; [E. O.] Beaman, Photographer; [Walter] Clement Powell, Assistent [sic]; [Frederick Samuel] Dellenbaugh, Artist; Frank Richardson, Barometrition; [Andrew J.] Hattan (General), Cook.

May 17

After breakfast the Major, Steward and myself climbed Fish butte, found some nice specimens of fossil Fish after a couple of hours search returned to camp. After dinner helped to finish boats.

May 18

At the boats again, covered them with canvas.

May 19

Painted canvas and made battings for cabins.

May 20

Had a row across the river, broke an oar on a gravel bank. Returned and had our pictures taken. *Emma Dean*, Maj. Powell, Jones, Helmsman, J. K. Hillers, Stroke, F. S. Dellenbaugh, Bow., *Nellie Powell*, Prof. Thompson, Helmsman, Bishop, Bow., Steward, Stroke, Richardson, *Super Cargo*, *Cañonita*, Beaman, Helmsman, Hatten, Stroke, Clem Powell, Bow.

May 21

Loaded boats and got every [thing] ready for a start. Major bought an arm chair which I strapped on the centre cabin for him to sit in in order to have a better view of the River as we go.

May 22

After a heavy breakfast which Mr. [Jacob] Fields, the trader at this place, had provided for us, we repared to our boats, coiled our ropes, and made everything ship shape. We took our oars, at 10 a.m. we shoved off from shore with three hearty cheers for the people of Green River City—(Its population consisting of about 100 persons, whites, mongolian, and copper collored.), which was returned from the same on shore. Down we went at the rate of four miles an hour, but had not gone far when the *Dean* struck a bank in the centre. Her crew got out and dragged it over. The *Nellie* and *Cañonita* followed suit. All got off safely. The Deacon broke his steering oar, the only damage done. Camped on the left bank for dinner and observation. Pulled out again at two, nothing of interest this after noon. Camped on the right bank in an old hut belonging to some old trapper.[1]

May 23

Rain this morning, wound up with a snow storm which left us under cover of the old hut enjoying the warmth of a huge fire. After dinner pulled out very cold and blustry. Camped on the left bank.

May 24

Woke up this morning with frost on our blankets. Pulled out again. Camped for dinner under the walls of a high cliff. Immediately after pulling out ran aground, jumped out and soon floated down merrily. Camped opposite Black Fork, a stream of about 75 feet, very muddy.

May 25

Pulled out at 7 a.m., ran along smoothly. Scenery fine. Camped for dinner on left bank. After dinner passed an island on which we saw a dear. Major fired but missed, landed, and got three. One swam the river and thereby escaped the fate of his comrades. Camped on the left bank.

May 26

Pulled out in the morning. Saw some Beaver. Major shot twice but missed. Camped on the right bank after a short run for the rest of the day. Jerked our venison.

Steward made geological sextions. Beaman took pictures. My friend Fred fixed his sketches.

May 27

This morning I was transferred to the *Nellie Powell* in place of Steward, who had not quite finished his work making sextions, while the Professor was anxious to make a certain point by 9 a.m. in order to get longitude and latitude. We had a hard pull of it but made it in time. We entered a mountain of Vermillion Red Sand Stone; it is called Flaming Gorge. Camped on the left bank in a beautiful cotton [wood] grove.

May 28

Water bright and clear. Major, Steward, and Richardson climbed the opposite bank to geologise the country. Richardson left them, for which he was nearly getting five days rations and an open road to Green River City. I fished, caught some very fine fish which we had for supper.

May 29

Pulled out again this morning. River snake-like.[2] After making several twists we came to one almost doubling and just at the centre we had our first rapid. Boats riding fine. Walls perpendicular, in some places, over hanging. Camped on the left bank for dinner, spent the rest of the day here. Prof., Steward, Bishop, Jones and Fred went across the river for Observation and Geological work. Major, Beaman, Clem and myself went up the river for pictures of the rapid and surrounding sceneries. Dropped down about 4 miles and camped near Kingfisher Creek on the right bank.

May 30

This morning the Major and myself climbed out, I for game, the Major for geological work.[3] Struck the Creek about 2 miles from camp. We tried the trout first, but they being so stubborn would not take our decoy flys, so we started for the hills. Climbed up some 2000 feet. No sign of dear but got a splendid view of Kingfisher Creek for some 20 miles we could see a serpent form coming down through Cotton Wood Groves. Got back to camp tired.

May 31

Had a good rest last night, pulled out early, passed through Kingfisher Cañon, ran a rapid, camped for dinner on the right bank. Beaman took some fine pictures of Kingfisher Creek and Cañon. Ran another rapid after dinner camped early on the right bank head of Red Cañon. Fred and myself at the oars, Hatten steering pulled Steward over on the opposite bank, the edge being thickly lined with willows, the boat got caught in a counter current flung her into the Willow brushing Steward off into the water, but we pulled him out in a saturated state.

June 1

Stayed in camp repairing and washing my unmentionables. While at work I can hear the dull roar of rappids, which means business tomorrow.

June 2

Got up early and immediately after breakfast we pulled out. After running a mile we came to the Rappids which had forwarned Business, and it was just "Biss" you bet. We got into a succession of "Raps" the boats half filling with water. At one place the river made a right angle turn, and just at the centre point a rappid. The *Dean* in running this almost filled, struck the rocky wall which tore out my oar lock. We would not have had this colision had it not been for the Major watching the *Nell* which was in a place which seemed disasterous. With only the use of two oars, we had hard work getting her ashore, as it was we had no time to spare for we were at the edge of another rappid. All hands jumped out to save the boat from striking on sunken rocks, pulled her ashore, and bailed out. The *Nell* struck a rock close ashore rolled over, the men jumping on a flat rock. Richardson being excited would have been taken under had it not been for Prof. who pulled him off. The boat as soon as she was relieved of her deckload of human freight righted up again. She stove her sides some, but not enough to hurt. Damage was repaired in 30 minutes. *Cañonita* came through all o.k. Passed two Creeks coming in on the right which we named Kettle and Compass, in commemoration of a kettle and compass we lost when the *Nell* turned over. Camped on the right bank for dinner. Beaman, Clem, Richardson and myself went up some ways to get a picture of the creeks, but could not get to it, the brush being too thick. Pulled out after dinner, ran several more rappids, and camped on right bank on the same spot where the Major had camped on the same day years ago.[4] Built a huge fire. Here we made our first portage, carrying the provisions for ¼ mile letting the boats down by line. After supper hung our clothing out to dry by the fire. Turned in early having done a big days work.

June 3

In camp this day. Fred caulking Bulkheads, and done his own washing for the first time in his life. The rest being all old soldiers know how to handle the "Kays of the Pianar" but I must give him credit, he done it well. Major, Beaman, Clem and myself took pictures, or rather Beaman took them, and we carryed the instruments.

June 4

This morning let the boats down by line 40 yards for this, when we got in to pull them through more rappids. After running several we came to one where we had to let down by line, after passing the rappid we got in to pull through more, in one of which Mr. Hook got drowned two years ago while attempting to follow the Major.[5] Camped for dinner on the left bank. After dinner concluded to stay all day. I muffled my oars with leather to keep them from slivering where the[y] work in the Locks.

June 5

Pulled out again this morning, ran more rappids but none very bad. Camped on the right bank for dinner. Prof., Beaman, Clem, Richardson and myself took some pictures of two Creeks coming in on the right, Francis & Cactus.[6] After dinner we pulled out with the *Dean*, looking out for Ashleys Fall, leaving the *Nell* and *Cañonita* behind, on account of barometrical observation and for Beaman to finish his pictures. We had not many miles to pull when we could hear the dull roar of the Fall and soon after we came in sight. Landed just above, unpacked our boat and waited the coming of the other boats which soon made their appearance. They soon followed suit unloading. Carryed our provision and baggage over the rocks first, then we got the *Dean* out of water and carryed her over, no easy task. Men worked hard and with a will. Her keel and sides were much injured by dragging her over the rocks. She being so badly scraped, the Major concluded to let the other two down by line, so we took the *Cañonita* next. We fastened a line to her stern and another to her bow. Down she went. When halfway over the fall we had to let go of her stern line in order to let her swing. While swinging around she got some heavy thumps on sunken rocks which filled her standing rooms with water, no other damage being done. The Major seeing how barely she escaped being knocked to pieces, concluded to try the first experiment with the *Nell*, which was done in about an hour. Camped for the night on the rocks. The Major named this Fall after an old hunter who had placed his name on a rock. fac semela ASHLey The Major tells us that we will have smooth water until we reach LaDore.

June 6

Started out this morning anticipating smooth water but had not gone far when we heard a noise resembling a rappid, but of course having been told that we would have smooth water, we thought nothing of it. But all of a sudden turning an angle we found that little rappid round the corner, but got through it all right. Down the river we went meeting rappids after rappids, some of them as swift as 15 miles an hour. At dinner some one of the boys asked the Major if we would have any more smooth water, when he answered "well about the same". We partook of our meal on the left bank. Before reaching dinner station, Major, Beaman, Clem, Richardson and myself went up a beautiful creek, took pictures of some falls. Major and Richardson left us to pull down some distance. Took Hatten out of the *Cañonita* to prepare dinner. When they got there, leaving me in his place, Mr. Beaman placed Clem on one side of the fall and myself on the other. Satisfied with our success we followed, we called this Snow Creek. After dinner ran some 10 miles of about "the same" as we had in the fornoon when all of a sudden we opened out into a beautiful park. Major called it Red Cañon Park. We camped on the right bank under two large Pine Trees. Took out our Ration to dry them. Caught a mess of fish.

| June 7 | In camp, drying ration. Washed and mended my clothing, half soled the Major's breeches. Only four of us in camp, rest went to climb a mountain some 4000 feet high. They returned at night, called it Mount Lena, after a bright flower of Cap Bishop's fancy.[7] |

June 8

Pulled out this morning. A little more of "the same" when we came to an open space. Major called it Browns Park. On the left hand we saw some herders. Landed and found that they had some mail for us. Camped for the rest of the day under a very peculiar shape Cotten tree. It resembled the horns of a deer. This park used to be called Browns Hole. The fact is, when a man "squats" on a piece of ground out here, his mud cabin is generally called his Hole. A man by the name of Brown, an old hunter, used to have his Hole here and hence its name. Several thousand cattle are kept here to fatten and then to be driven to California for consumption. This part is a splendid grazing country that is along the river bank.

June 9

I was surprised this morning to find that Richardson was to leave the Expedition.[8] The Major says that he is not strong enough to stand the trip, so he will leave us the day after tomorrow with Mr. Harroll [Harrell] for Green River City. Some cattle were drove up from which one was selected for our use. We gave them flour and sow belly in return. A greaser lassood one, a nice young heiver about 3 year old, made a rather bungling job. I should advise him to go to San Francisco and there hire with the dog catchers. They would soon teach him how to throw a Lasso.

June 10

Wrote two letter[s], one for each brother in Frisco. Mr. Harroll will mail them at Green River. He ate supper with us, a Texan, and quite a gentleman.

June 11

Everybody had his little tricks ready. Mr. Harroll provided a fine riding animal for Frank R. and a pack mule to carry some nagitives for Mr. Beaman to be sent to Mrs. Powell at Salt Lake. Everything being ready the boys all shuk hands with Frank. He felt bad about leaving, the tears were in his eye but failed to shed any. I felt bad about him leaving. Immediately after we shot out like an arrow. Ran a few rappids. Low grounds for a few miles on each side, only here and there where the River had cut through some small hill, could hardly call them cañons being used to such large ones. In starting Fred broke one of his oars. Camped on the right bank in Swallow Cañon on a narrow ledge for dinner. The River through this cañon is very quiet. Walls about from 500 feet to 800. Pulled out at 2 p.m. when we opened out again into the Valley. Ran down to a beautiful Cotton Wood Grove on the right bank where we remained to jerk our beef.

June 12

In camp all day mending and washing. Major, Steward and Jones went accross the River to get Geology and Topography. Beaman took picture of camp.

June 13

Pulled out early this morning. About a mile down we met Mr. Bacon and a Georgean Negroe. Mr. Bacon is a Brother-in-Law of Mr. Harroll. We had promised to ferry him accross to some cattle he had on the other side of the river. The water being up high he did not like to swim his horses, and therefore did not go over. Bid him goodbye and pulled out. Lashed boats together. River being very wide and deep we floated down while the Major read aloud the Lady of the Lake. On the left came in a Stream called the Vermilion, so called by Fremont. A stream of about 75 yard in width it starts about the Park, at the present it contains but little water. Camped on the left bank for dinner. After dinner dropped down to the head of the Cañon of Lodore where we camped in a Box elder, or Aconigundo Grove. Musquitoes were so thick in the bottom that we had to move upon a bluff if we desired to keep all our blood. River still rising, bully for us.

June 14

In camp. Fixed the Major's Moccasins. Washed my clothing. Our soap is getting short. Fred and Steward went out this morning with two day ration, to geologise and sketch some distant mountains. Waiting here for Mr. Harroll to return from Green River with mail which had been sent to Salt Lake after our departure from Green River, for which he will telegraph to Mrs. Powell, who collects all of our mail. After dinner climbed mountain on left of Lodore, Prof., Major, Clem and myself.

June 15

Still waiting. I planted our flag on a high bluff overlooking the Valley below, as a guide for him to our camp, wondering if ever our flag had kissed the breeze from these lofty mountains. My friend Fred and Steward returned about 1 p.m. tired and hungry, having cached their provision the first day, but being unable to find them again had to do without until they made camp.

June 16

Still waiting in the evening our old friend Mr. Bacon rode into camp. Informed us that the mail carryer would be in late in the night or early next morning. He had supper with us but declined to share our hospitible ruff or that portion of the Canopy which covered our bluff.

June 17

Our mail carryer did not arrive till near noon when he brought only two letters, one for the Major, the other for Prof. The rest of our mail had been sent on to Uinta. Pulled out after dinner and entered the Rocky Gate of Lodore,[9] some Newspaper correspondent has said: he who enters here leaves all hope behind. Walls about

2500 feet, as we entered we could hear the roar of rappids. Ran two in fine style, shipped but little water. Beaman took pictures of rappids while the *Dean* went on alone for a few miles farther down but soon stopped and camped on the right bank a little above a roaring rappid nice music to go to sleep by.

June 18

Beaman took couple of negatives [of] a Grotto in the cliff. Steward called it Winnies Grotto.[10] Clear cool water running down the walls. Pulled out and ran that roaring rappid whose noise had lulled us to sleep. Done it handsomely. Shipped but little water. We watched the others come through as they had done from the shore. To see one of the boats run a rappid they resemble the bounding deer through a forest of fallen trees. Ran another bad rappid. Boats nearly filling, plopping through the waves. Pulled hard to save ourselves from going over another before barking boats. Camped on the right bank for dinner. Ran two more rappids and then camped for the night ahead of the Falls where two years ago the "No Name" was lost. Major calls it Disaster Falls. Beaman narrowly escaped going over. All hands stood aghast for a moment, five feet more and he would have shared the fate of the No Name.

June 19

Ported our provision over a rocky hill, then let the boats down by line a distance of ⅛ of a mile. Loaded up and let them down to the next fall. At this fall we found a sack of flour which had been rescued from the No Name two years before. Andy baked biscuits for dinner out of the flour. Found it in perfect condition with the exception of a crust of an inch thick on the outside of the flour. Unloaded boats again. Ported provision for ½ mile over a hill rocky all the way. Let one boat down by line after lifting it over rocks. Men being completely tuckered out Major concluded to leave the others till morning.

June 20

Let the other two boats down. Loaded up and pulled out. Let the boats down by line around lower Disaster Falls. Had dinner among a lot of sage and cedar, made a short portage. Ran out quite a ways. Stopped just above Cascade Creek. Pulled out immediately after. Ran a bad rappid. Ran a little to close on an island. Got five very bad raps on our keel. Camped opposite a cliff 2800 feet high, called the Cliff of the Harp.[11] From this Camp a very peculiar mountain is visible. It is in the shape of a wheat stack, and therefore called so.

June 21

Let the boats down by line over a rapid found a vice and axe lost by party 2 years ago. Camped for dinner on the right bank, close to the waters edge. Ran a rushing rapid immediately after dinner. Two feet more to the right of the course we were running and the *Emma Dean* would have been shivered to piece[s] [on] a large rock projected from the wall. The chanel lead close passed it and the suction carryed us closer than desired. Ran down a few miles of rapids until we came to one which stopped our rushing speed. Landed and let boats down by line. Pulled out for half a mile, stopped before another "Rusher." Let down by line. Had hard work to clear away an old Cotton Wood Tree which obstructed our passage. When about to leave Fred missed his sketches and found that he had left them at the second Portage so he had to climb a steep cliff to get to them. We dropped down a little way where we waited his return and then pulled down to Head of Tripplet [Triplet] Falls.

<table>
<tr><td>*June 22*</td><td>In camp Major, Prof. and Cap climbed Dunns Cliff,[12] part of the Sierra Escalanta,[13] 2800. Washed my clothes and caught a mess of fish.</td></tr>
</table>

June 22

In camp Major, Prof. and Cap climbed Dunns Cliff,[12] part of the Sierra Escalanta,[13] 2800. Washed my clothes and caught a mess of fish.

June 23

Unloaded the boats and let them down by line, ported provision a short distance over the rocks. Loaded again and let them down by line about a quarter of a mile, then unloaded, let the boats through a rapid and ported provisions some distance over a hill. Had dinner at the end of the portage and pulled out. Ran four rapids when we came to one whose noise we had heard for some time above the roar of all the others. Landed at the head of it, unloaded the *Dean* and let her down. As she was going through the shoot she struck on a rock and keeled over. Pulled her up, righted her and found but little water in her forward cabin, no damage done. Let her down some ways, ported provisions, loaded her and then took her down the stream by land the water was so shallow near the shore that we had to slide her along on the bottom for a half of a mile. The river is filled with large rocks with a fall of some twenty feet. The water rushes through these makes it one sheet of foam. We let the boats down for some distance when we unloaded her again and let her down for ¼ mile by line, ported provisions, climbing up a steep bank 50 feet high. Reloaded and then got in to pull her some distance to a little sandy beach. The quickest ride I have ever had in a boat I had while going this short distance. Made her fast and then returned to camp tired, but oh how hungry. Major very appropriately called it "Hells half mile".

June 24

Brought down the other two boats and got through just in time for dinner on the beach. The sage and cedar at our breakfast camp caught fire and the dense smoke rolled up to the peaks some 2800 feet in height and then out of "Lodore". After dinner pulled out, ran three tolerable rapids, and camped at the head of another. The *Nelly* in coming over one she ran on a rock, poised amid air and for a moment was doubtful of her fate, but a wave came and took her off nearly capsizing.

June 25

Fred, Steward, and I helped Beaman and Clem carry their instruments up the mountain from which Leaping Brook comes hopping down to the river. I attempted to sketch some of the falls, but failed to do justice to its beauty so gave it up in disgust. After dinner let boats down by line. Had to lift them over the rocks and then again some ways by line. Got in and ran some bad rapids. The *Dean* shipped her standing room half full, the *Cañonita* ran her bow on a rock but the current took her stern around and so slid off without damage. Reached Alcove Park about 4 p.m. Beaman took some pictures of a Creek of that name and its surrounding beauties. At this place two years ago the last Expedition after having been landed a short time, their camp caught fire, in which they lost most of their mess kit and some clothing. They had to "Git" to save themselves. Ran down to the mouth of Bear or Yampa River where we arrived at seven o'clock.[14] Major says will remain some days.

June 26

In camp fixing up for a trip up the River tomorrow. Loaded the boat with three days provision. Caught some fish.

June 27

Major, Jones, Beaman, Hatten and myself started out up the Yampa, Hatten and myself on the Oars. Made good headway for a few miles when the Current began to tell on us, until finally we could make no headway. We got out with our line and imitated the mules on the Towpath. Camped on the left bank for our dinner. Beaman took pictures. After dinner we had something more than strong currents to contend against for we heard the noise of a small rapid "around the corner". All hands got hold of the rope, Hatten and myself leading some ways. The shore being thickly fringed with willows made it an awkward thing for us to get her over a fall of about 3 feet. Major proposed to get her over stern first. Hatten and myself went back to the boat to help lift her over some rocks. Got her over all right, Major, Jones and Beaman holding on while Hatten and myself went to the end of the line to hold on while they came to our assistance. They had hardly left the boat when the boat bulged and down she went but fortunately was stopped by Hatten. The rope had got around his leg which was fast between the rocks and myself holding on like grim death to a negroe, we stopped her half way over. The rest of the party coming to the rescue soon brought her up when a second time she got away but stopped herself by bumping against a rock. Fortunately for Hatten he received no injuries. Got in and rowed a little ways, then had to get out again to tow it, so by alternate rowing and towing we got up some 4 ⅛. Camped on the right bank. Beaman tried to take a picture of Cañon with moon in but failed to get the moon. Walls from 500 to 1000, mostly lime and sandstone. Major called this Yampa Cañon and the little park Grisley Park on account of the numerous tracks we saw of bruin, but failed to see his Majesty himself.

June 28

Major and Jones climbed out but failed to get high enough for Topographical observations. Had dinner. So we harnessed ourselves up again and towed and pulled up a rappid, after passing this we got in and pulled for some distance. Saw some mountain sheep, tried our guns but failed to hit. Little farther up we saw seven high up on a hill looking down on us. Camped on the right bank on a narrow ledge under a box elder. Major and myself had our bed to slanting and so kept sliding down the "Cellar door" all night.

June 29

Pulled out early, ran up a short distance and then got out to tow. Camped on the right bank about 10, had a lunch. Major climbed out but failed to get high enough. Returned and reported some fine views from the top of mountain, so we took Beaman's instruments and all started up. Returned about 1 p.m. Started out with the tow line and whenever we had a chance we pulled. Major shot at some geese but failed to hit. We tried but followed his example. Then we concluded that they were "Wormy, Story of Fox and Grapes." About four p.m. we landed on the left bank opposite an island. River taking a sharp bend to the left looking up three latteral Cañon[s] coming in here, in one of which is a creek of pure sparkling water. The "General" Hatten went up and explored it. A little farther up a beautiful park opens out. Major called it after him and so it goes down on the map as Hatten's Park and Hattens Creek. Found our rations short, only enough for breakfast. I caught a mess of fish in lieu of bacon.

June 30

Passed a very cold night, the coldest on the trip. Major "spooned" up like a good fellow. Got up very early. Helped Hatten get breakfast ready. At sunrise Major and Jones started to climb a bald mountain some two miles from camp. Beaman, Hatten and myself dropped down stream to a latteral Cañon for some picture on the right. Stopped at the head of a rapid, fastened the boat, carried the instruments some distance down the stream. Hatten and myself returned to the boat, rowed up some ways and then crossed to the left bank to wait for Jones and Major and to prepare dinner, but I had to catch it out of the river first. No coffee or tea, no bacon. All we had was one biscuit and some sugar. I was lucky enough to catch five large fish, one apiece, which we roasted. About one o'clock the Major and Jones returned, having been very successful in getting observations. I gave them one biscuit and one fish apiece and all the water they wanted to drink—but Jones was sick of fish, he begged me to take them out of sight. This finished, we dropped down to where Beaman was waiting for us. Issued him his ration of one biscuit and fish which he ate while Hatten and myself pulled for our supper on the Green as best we knew how. Made three portages by line and lifted boat over rocks twice. Made all told 12 miles. Found Clem and my friend Fred in camp in charge of culinary department, the rest having climbed a mountain, who soon after returned and that evening was spent in telling adventures.

July 1

Rested and cleaned up.

July 2

Made Major a [pair] of moccasins and mended my shoes.

July 3

Started this morning from Echo Park as the Major called it on account of the great echo, there being a perfect wall of solid sandstone some 600 feet in height it runs some 500 yard along the river at a thickness of 200 feet at the base and quite sharp on top [Steamboat Rock]. The River turns back on the other side, making a sharp bend round the point of the wall. Left the *Cañonita* behind. Beaman wanted to take some pictures immediately after making this bend we entered Whirlpool Cañon. At the entrance we were saluted by the roar of a rapid which we ran. Walls 500 feet and sloping from the river to 2500 feet in hight. Passed over two rapids, then came to one which we thought best not to run. Waited for the *Cañonita* but did not come in sight until we had the *Dean* and *Nell* let down by line. Had dinner on the right bank opposite Rapid. Let *Cañonita* down and started. While at Echo Park water fell 3 feet which helped us considerable going through this cañon. The river is not so swift and therefore less whirlpools. The whirlpools are occassioned by the rush of water against the ragged rocks which project from the wall. Nearly the whole wall presents the appearance of a stair laid down on its side. Saw some 9 or 10 sheep on the right bank high up on a Tailess [talus]. Jones and Fred started up to cut them off but they were too late. Camped on the right bank at the mouth of Brush Creek,[15] so named by Fremont, a nice little stream of pure water, full of trout. I started up for a mess but they were too stubborn to take my grasshopper.

July 4

Oh glorious Fourth. The boys saluted the sun with one shot for every state in the Union. Major gave us a Holyday, but Prof, Jones and Hattan climbed out and did not return until night. I being almost barefooted thought of a plan to make myself a pair of "Wooden Shoes". I succeeded. Made the soles and sides of wood and covered the top with canvas. I congratulated myself, but low, while walking over the rocks the bottom of one split, so my whole days work had been in vain. At 6 p.m.

we had dinner, but such a dinner not to be sneesed at in this wild country. I had procured some canned fruit while at Green River, which I had stowed away in the forward cabin so that no one knew except Fred, he had a few pounds of candy which he had brought, Hattan being away so Fred cooked dinner. Had a lunch about 12 n. No one suspected what we had for dinner, everybody being surprised to find strawberries, peaches, tomatoes, pies and candies, ham and a beautiful strong cup of tea. Everybody ate hearty and enjoyed it more and better than they would have done at home. After dinner lit our pipes and cursed the Irish until it was time to seek the fond embrace of Morpheus, and so ended the glorious Fourth of July of 1871.

July 5

Pulled out this morning. Ran some half dozen rapids, bumped on rocks here and there but not serious. Camped on left bank under a cottonwood tree for dinner. Jones, Bishop and myself went across the river to shoot some sheep we saw on the hill but failed. After dinner pulled out, ran a few more rappids, at one place barely saved ourself from being crushed to pieces by rocks. We ran to[o] close not expecting such strong suction toward them. Ran one more, an old monster and then opened out into a park, the little low hills being highly colored and its different shades were as numerous as that of the Rainbow, in Horizontal stripes. Major called this Island Park. The river breaks up into small streams as it runs through this low country, forming numerous little Islands. The Park is about 8 miles long by River. Camped on the right bank in a Box Elder Grove, the Head of Craggy Cañon [Split Mountain Canyon].

July 6

In camp, fishing and fixing up my traps. Major, Prof., and Steward went out for fossils, returned at noon brought in a few, went out again after dinner, and returned about 4 p.m. with quite a lot. Major asked me how far I could walk in a day and I was surprised. I answered carelessly. Oh about 30 miles. When he asked me if I could not make 40 I told him that I might at a pinch. Then he said that he had notion to go to the Uintah Agency where our rations are stored, it being 60 miles but we had to make 40 miles the first day in order to reach water. Talked it over with Prof. who persuaded him to take boat and go as far as the Uintah [Duchesne] from there foot it, so he finally concluded to go by boat. Made arrangements accordingly. Exchanged Fred for Capt. Bishop.

July 7

Got up early and pulled out at sunrise. Had not gone far before we struck a rapid which we ran, little farther down still another, made a portage by lifting the boat over the rocks, worked about 40 minutes, had hard work, pulled out a little ways and then let down by line. Rapids all the way, got in pulled across to the right bank, made a portage by line. Boat got stuck on a rock, Jones went in the water a little ways to push her. He gave her push when all of a sudden she slid off. Jones losing his hold went headlong in the River. All I could see of him was his white hat. Fortunately for him he got hold of a rock, thereby saving a swim down the rapid and perhaps his life, Having left the other two boats behind to finish up getting Topography and Geology. They will spend three days in this cañon for that work and also for Beaman to get pictures. After letting the boat down we got in to pull but soon were stopped again by rapid. River breaking up into two streams, forming an Island in the centre. Landed on this. Took the right hand chanel, let the boat down by line. At the foot of the Island got in for a short distance ran a rapid and camped on the right bank for dinner. Immediately after pulled out, ran a rapid, then led down by line. Pulled a little ways, then had to get out and let her down again by line. Ran two more, then came to one a huge old monster. Let down again by line. At this point we could see the end of the cañon. At the end we mistook the chanel and got into the wrong one. Had to get out to lift the boat over the rocks. Then rushed down the river at 10 knots an hour, could fairly see it go down hill. Passed over some small rapids, all o.k. Then we opened out into the Valley of the Uintah. Craggy Cañon is formed of a mountain split in two by the River. Major first called it split mountain, but changed it to Craggy. Innumerable crags, peaks and Pinnacles can be seen. Ridges running perpendicular to the River. The scenery is grand. Some peaks ran up to nearly 3000 feet. Those huge old crags bespeak age, a fine study for Geologists, but I decline to puzzle my brain with the age of a stone. Got out of the cañon about 4 p.m., having made seven miles, pulled in seemingly smooth water for seven miles more then camped on the left bank under some cottonwood, near a lot of Indian "Wickiups", Wigwams. Ran 19 miles, killed three geese but on getting the feathers off them we found them to[o] poor to eat.

July 8

Got up early and after breakfast pulled out, leaving our geese on the bank for wolves and coyottes. Here the river bends and twists through the Valley very wide and shallow. Had to get out to push boat over bars several times. Camped on the right bank for dinner. Immediately after dinner pulled out again. Rained and blew fearfully, mixed with peals of thunder. Camped on the right bank for the night. Shot some geese in the afternoon which I picked for the feathers to make myself a pillow. Found one to be pretty good. I cleaned it for breakfast. Plenty of mosquitos here, could hardly sleep for the rascals.

July 9

Got up early. Prepared my goose and some ham while Capt. Bishop baked the biscuits and Jones brought food for fire. I found my goose too tough for my ivories, so left it for the wild beasts. Pulled out while the Major read about Harrold the Dauntless, one of Scott's poems, hardly any current, had to pull her through in order to make headway. While passing near a little knoll we saw two antelopes, one drinking, the other on the hill coming down. Dropped down easy and then landed below the hill. Major got out with gun had a good sight for killing them. He cocked his gun, aimed, fired, but low it missed. The noise frightened them and away they bounded. On examination found the shell of an exploded cartridge in the Bore, in place of a primed one. Moral, before you shoot at wild game be sure that your gun is loaded or 10 to one you miss getting it. Passed Powells Lake on the right. Major mistook the outlet first for the mouth of the Uintah but soon found his mistake. This lake has been called after his Brother, Capt. Morris [sic, Walter H.] Powell who two years ago was with the Major on the former expedition. It is some two miles wide and about 3 or 4 in length and contains very clear water. Camped on the left bank for dinner. Pulled out as soon as we had our chuck down. Ran until 4 p.m. when the Major pointed out a large cottonwood Tree as our destination, which we reached a little afterwards. Found a house on the left bank, pulled over and found it vacated. Major thinks it belongs to a man by the name of [Pardyn] Dodds, the former agent for the Uinta Utes Indians. Pulled to the other side again where we camped for the night. The Uintah River joins the Green about ⅛ of mile farther down. At our Camp is the Old Overland Stage Companys Ferry. The Pioneers Wagon tracks are still visible. The boat which was intended as a Ferry boat is sunk here on the bank, she having been hauled close to shore and filled with sand to sink her, two stakes being driven in to keep her from going down stream in case of the water washing the sand out, but no danger of that. Where once was the edge of the River is now almost highland excepting high water mark. This road has never been used. Only one train went over it with the constructing party. It runs from Denver to Salt Lake in an almost straight line, almost from East to west.

July 10[16] Got up very early, packed the Major's blankets and mine into a little knapsack, got a days ration into a bag, had breakfast about five. Major equiped himself with the ration and a Henry Rifle, while I being an old soldier slung knapsacks. About 6 we started out for the Uintah Agency. Made 12 miles in four hours. Where we crossed the Uintah River had a wash and a smoke. Forded and struck out. Made some 8 miles more when we came to the River again. Here we concluded we would have a cup of coffee. I soon had it made. Drank and smoked, then up and off again until about 6 p.m. when we struck water again. Had a cup of coffee and smoke, when we crossed several branches. Walked until nine p.m. when we came to a crossing where we lost the trail in the woods. All our efforts to find it proved fruitless so we concluded to camp out and walk in early in the morning. Laid our few blankets on the ground and turned in. Slept sound.

Got up early, cooked our coffee, smoked and started off soon finding our trail. Walked a half an hour when we saw the Agency at whose breakfast table we found ourselves soon after doing justice to a hearty meal. Found lots of mail for the boys, one for me from Dick. People have called it 45 miles but Major called it about 40. Found the Lord of the Manor gone, had started three days before our arrival to Salt Lake City. A Mr. [J. J.] Critchlow by name is the Agent for the Uintah Ute Indians, a sort of second hand Prespitarian Minister, and as I understand from some of his men, he is more Mulish than rightious. Before leaving he issued orders to his foreman that under no circumstances he was to let Major Powell have teams to carry our rations down. It took us aback. We did not know what to do, but Mr. Besser [George Basor] the Trader came to our rescue. He offered us his team. Now supposing no teams were to be had, would we pack our rations down 45 miles on our backs? I think not. I would have harnessed up one of his teams without permission and drove it down to Green River and then returned. I don't think one of his men would have resisted my taking the team. The men say when he took charge he wrote out regulations and stuck them up at the Carpinter and Blacksmiths Shop. The first thing he done was to abolish the ten hour system, men had to work from sunrise to sundown, one hour for dinner. All Government work according to the latest law is only for 8 hours during the day, extra pay being paid for overtime. On the whole I dont think much of Critchlow taking the word of his men, and his regulations confirmes them. Found eight buildings and lumber for another. One office, one Blacksmiths and Carpinter shop, one for mens quarters, one for kitchen, one for stores, and one barn. Mr. Besser and Mr. Dodds each own a house. Mr. Dodds has a large drove of cattle in this Valley, but talks of going farther down the River, His Royal Highness, the King of the Valley having hinted that he must take his stock away, that the Valley belonged to the Indian Reservation. Good many [of] the Indians came to pay [the] "Americats" a visit expecting of course heaps of "shug" and "frour" in return. Smoked the Pipe of Peace with Old Tuckanoana,[17] a fine looking man I should think about 45 snows old and 6 feet in his stockings. His father used to be chief of this tribe, but he declined, he went to farming, the only one who seems to go at it with a will. He has got a nice little farm fenced in. Talbia,[18] a fine looking man, is chief at present about the same age, married or rather took unto himself a nice looking Squaw of about 16 snows, the best featured woman I have seen among the whole tribe. The Squaws do all the work while the head of the family takes it cool by picking the lice off himself which abound in his dirty carcass. Whatever they put on it stays there until it rots off. They never think of taking it off to wash. Their breeches are of Buckskin and are sewn on their legs skin tight, with sinnew or "Tammu" as they call it. The women look dirty, some wear Buckskin dresses, fixed with beads. A dress of this kind is valued at about 50 dollars. They carry their children on a piece of bord with a basket on top as a cover for rain or sunshine. The "Pappoose" is tied to the bord and slung on the Squaws back. It is very seldom that you heard one cry, and they are called pappooses until they get married. Major took possession of Mr. Critchlows Office, in which we slept.

July 12

Had a good nights rest. Got up, had breakfast at an early hour. Major talked of going to Salt Lake City. Mr. Dodds offered his horses and thought of going himself. Had a look at our rations, found them in good order, but no salt or soap, articles we were very much in need of, good many other articles which had not been sent from Camp Douglass, U.[tah] Territory]. The most I missed myself was ham. The Major had bought and paid for these things and why they did not send them is more than I can tell. Made an exchange of bacon and sugar with Mr. Laighton, [Thomas Layton] the man in charge for some soap enough to last us for two months. Also took half of a hide for soles of moccasins. More Indians today, gave them a sack of flour to devide among themselves. How they did it I cannot tell, but I think it remained with the two aforenamed Indians to whom it had been given for distribution. Major concluded to go to Salt Lake City tomorrow morning, leaving word and letters for Mr. Thompson how to act. Got everything ready for his journey tomorrow.

July 13

Major and Mr. Dodds left early this morning. After breakfast I went down in the meadow for the horses, wanting to get Mr. Camble's [J. L. Campbell] horse shod for me to ride down with to the boats, and let Mr. Thompson ride up who was expected to be at the crossing about this time. As I got back with the horses I met Capt. and Jones. I was glad to see them. They had come up the Uintah some distance with the Boat and there cached it under some willows. They were a tired pair when they got here. They too had lost the trail and camped on the same ground on which the Major and myself had camped on before. Shoeing Camble's Horse, I ran her for some five miles as hard as she could run, to take some of the fire out of him, which made my thies very sore not being used to riding. Spent the rest of the day talking and resting.

July 14

Cap, Jones and Mr. Layton went down to the Indian farm some four miles from here. They returned at noon. I made a pair of moccasins for Mr. Black, one of the farm hands, in whom I recognized an Old Brother Soldier having served in one Brigade together, with Butler in front of Richmond. Had quite a chat about Old Times Rocks. Three Indians belonging to the White River Reservation came riding past the Agency about 9 p.m. I had just "turned in" near a Haystack when I heard three shots fired. I thought some of the boys were discharging their pieces though I heard the whiz of the bullet when immediately after I heard Jones call me. I answered but did not get up, thinking that he was coming to bed and wanted to know if I had retired, when a second time he yeld like a stuck bull for me to come up. I went up to where I saw them all standing, was surprised to find them all armed. Of course I got my Henry XVII and joined them to slay anything that might [be] hostile to my life or good for my stomach. Mr. Besser the Trader soon joined us and said that the Indians who had fired had asked him if he was friendly to Indians. When answering in the affirmative they rode on. Little while after we heard the War whoop, which sounded like business. We remained on our guard until 10:30 when we turned in after barring everything.[19]

July 15

Heard this morning that the three Indians who fired were excited about a murder that had been committed near White River, and 10 had died from some kind of epedimic, the murderer had escaped, belonging to the same tribe. Capt. and Jones left this morning saving me a ride down to the boats on horseback. Fixed my shoes and made myself a pair of moccasins.

July 16

Helped to unload hay to keep myself busy. Expected Prof. and Beaman up but did not come.

July 17

Helped the Blacksmith a Mr. [Martin] Morgan from San Francisco to weld out irons for the keels of our boats, which needed them very bad. A mistake of the Major, thinking that it would make them too heavy to have the keel ironed. In going over the rocks the wood was almost worn even with the planks so it was necessary to have irons put on. No Prof. this day.

July 18

Finished the irons and read Scribbner's Magazine. Prof. and Beaman came up about 7 p.m.

July 19

Got our rations ready for tommorrows start. Helped Mr. Besser fix up his wagon and got my own traps ready.

July 20

Had breakfast, loaded our rations, got started about 8 a.m. leaving Prof. and Beaman behind to photograph some of the Indians and the Agency, or such views as they might find of worth. Besser and myself started slowly, day being very warm and no breeze. Camped for the night at the Uintah crossing. Fastned two horses with ropes while we hobbled the other horse and the mule, having crossed before we camped, thought them secure. About 12 at night I was awakened by plunges in the River. Thinking that they had gone down to drink waited for some time, but they did not return, so I concluded they had crossed, but could hardly satisfy myself about the horse and mule crossing they being hobbled, and how they could cross a rapid was more than I could account for, but I could see no trace of them so I crossed. Ran through the woods when I laid down to listen, but could not hear them. I ran half mile farther when I listned again, when I thought I heard something bound. I started and found them making their way back. I got hold of one rope, mounted him and drove the rest back. They had gone two miles. I had a fine shirt tail Parade, but no one to review me. Had them picketed afterward.

July 21   Started out at sunrise. After going up one hill and down another we at last reached Green River about noon. Shouted for the boys who soon made their appearance on the other side. All hands came over to welcome me back an[d] indeed I was glad to be "at Home" with the boys. Mr. Besser had dinner with us and started back about 2 p.m. Rested the rest of the day.

July 22   Washed my clothing and done a little mending.

July 23   Made the Major a pair of moccasins and soled his old pair.

July 24   Having made these so well, everybody wanted a pair, and being all barefooted I made a pair for the whole outfit. Commenced and finished Capt's. Fred, Steward and Cap started up White River took three days rations.

July 25   Saw some antelope this morning. Started out for them but they saw us before we got within range.

July 26   Took the *Dean* out of the water, put iron on her keel and caulked and pitched her.

July 27   Finished the last pair of moccasins. Sawed slats for cabins. Fred Steward and Cap returned at night having gone up 50 miles both floated down on a raft.

July 28   Washed and fished.

| July 29 | Went out hunting but could not see a single game. Prof., Beaman and an Indian returned from Uintah. They had horses from the Agency which the Indian was to take back. Major had gone to the mouth of Dirty Devil [River]. |

July 30

The Indian started back this morning.

July 31

Commenced packing our rations in the boats. An Indian made his appearance on the opposite bank. Went over and found a son of nature and his Squaw. He proved to be the son of Douglass, the Chief of the White River Indians. This young hero was on an elopement tour, having run off with another chiefs daughter, she being another bucks promised bride. These two showed open affection for each other, the only instance of the kind I have ever noticed among them.

August 1

Beaman took a photograph of these two children of nature. Had no objection to have their picture taken.

August 2

Prof took latitude and time while the rest worked at the boats.

August 3

Our Indians paid us another visit. The Squaw made Jones a pair of moccasins but did not make them to fit good. I went over where she made them but dropped her work as I approached. Asked if she had enough of buckskin taken she said she had. I thought she had taken enough for two pair so I overhauled her storage and found enough stuff to make herself a pair, but I took it out of her bag. She only laughed. These Indians will steal and when caught only laugh.

August 4

Made everything ready for tommorrows start. Our Indians bid us goodby. He started down the River. We christned him Louchinvar.

August 5

The long looked for day had come at last and at 8 a.m. we pushed off. *Nell* leading, *Dean* next, and the *Cañonita* bringing up the rear. Passed the Uintah about 50 yards wide on the right, the proper name for the Uintah here is DuSaine [Duchesne] or at least the Indians call it so. The Uintah joins the DuSaine about 12 miles from its mouth, a stream not half the size of the DuSaine, and why the latter should lose its name I cannot tell. The DuSaine starts at the Wasatch Mountains and the Uintah from the mountains from which it has its name. The two combined supply the Green with considerable water. About a mile below on the left we passed the "White," a stream about the size of the Uintah. It starts from the Parks in Colorado. Made 10 miles in the forenoon. Beaman killed a beaver but failed to get him, he sunk. A little farther down we met our Old Friend Louchinvar and his other greasy half who was jerking some venison. We exchanged a few pounds of "shug" for some Jerk and pulled out for a few miles. Then camped on the right for dinner and observation. Pulled out again at 2:15. River very quiet and full of sand bars. Got out several times to push boat over. Ran 9 miles this afternoon. Camped on an Island. Called it Beaver Island. Called it Beaver Island on account of the many here.

August 6

Concluded to stay in camp this being Sunday. Prof., Steward, Jones and myself climbed out on the left. We left the valley about 3 p.m. yesterday and are gradually running into what is known as the Little Mountains by the people of this country, but not laid down on the map. Looking from the top it is the most dessolate looking country I have ever seen. Even the hardy sage dies for the want of nurishment. The only sign of life is on the bank of the River wherever it is sandy the willow and cottonwood grow.

August 7

Pulled out this morning early. River still very quiet and full of bars. Got out and in most of the time to push boat over. Shot a beaver but failed to get him. He slipped out of my hands and sunk. Camped on the right bank for dinner. Prof., Steward and Cap climbed out some 1050 feet for observation. Started at 3 p.m. Dropped slowly down. About five o'clock Prof. saw a beaver on the bank which he killed, rolled down to the waters edge where he wollowed in the mud. Steward jumped out but before he got him into the boat he was covered with mud from head to foot. Camped on an island. Bish and Clem skinned him and so in place of oxtail we will have beavertail soup for dinner tommorow. Passed a weeping willow tree on the right, first I have seen on the trip. Made 13 and ⅗ miles this day.

August 8

Had beaver steak for breakfast. Pulled out at 8 a.m. Ran a few miles. Camped on the left bank. Boys got sick of beaver so in place of beavertail soup we had bean soup for dinner. Beaman got some fine pictures of a latteral cañon. Pulled out after dinner for about a mile and a half. Beaman saw some nice views so we stopped. Ran a rapid about a half mile after leaving dinner camp which we called the commencement of the Cañon of Dessolation. Camped on the left bank for the night. 3 and ⅗ miles.

August 9

Camp this day. Prof, Beaman, Clem, Fred and myself climbed a small mountain for to get a photograph of an amphitheatre, but failed to get it on account of the darkness of the day. Left the instruments on the mountain for tomorrow. Spent the rest of the day washing my clothes. Walls opposite camp 800.

August 10

Prof, Beaman, Jones, Clem and Fred started up the mountain on which we had left the instruments. Being a fine day they got a beautiful picture. The river doubles up at this point and runs back within 200 yards of itself, for nearly a mile a narrow wall dividing it. Party returned at 11 a.m. Cap. and myself cleaned and filled barometer. Had dinner and pulled out about 2 p.m. Stopped for observation on the right. Beaman took some pictures of an alcove. Heard the noise of a rapid once more which we ran. The *Dean* struck a rock in going over but no damage. Camped on the left bank behind a line of Box Elders. Got out once this day to push boat over sand bars. Made 10 miles this afternoon.

August 11

Prof, Bishop and Steward climbed out this morning for observation. Returned about dinner, while we dropped down about an ⅛ of mile. Andy and myself got dinner while Beaman took pictures. After dinner pulled out, ran down for about 3 miles when we struck a nest of rapids. Ran the first one in fine style. The second looked better than the first, but as the *Nell* went over, struck a rock, but slid off. Jones, Fred and myself with the *Dean* struck the same rock, but being a little heavier loaded, our stern hung on. We pushed a while but to no purpose, so I jumped out at the stern, obtained a foothold on the rock, lifted her up, and off she slid. The *Cañonita* struck a rock about the same time farther up which brought her broadside to. In swinging around she struck another which broke two ribs, leaked very bad, and had to haul her on the beach to brace her side. Delay one hour and forty minutes. Just below us was another which we ran all OK, making three rapids in one-half mile. Ran two more. Struck rocks in both, being very shallow. Got out to ease her load. She went over all right. Camped on the left bank in a cotton wood grove. Afternoon work five rapids and made 6 miles. Boys shot at some beavers but failed to hit.

August 12

Started out this morning early with a rapid immediately after starting. All the boats stranded on it. Got out and pushed them over, being very shallow. Ran along for a mile when we came to a lot of rapids. Ran them all. Had to get out in two to push boat over. Saw a deer drinking. Landed and tried to cut off his retreat but he escaped by a gulch. Ran 8 rapids and made 5 ⅞ miles. Camped on the left bank in a cotton wood grove for dinner. Concluded to stay all day. Fred and myself climbed out for some pine pitch. We had a hard time getting up, being very steep and shaly. While climbing along a tailess I got hold of a rock for support which gave way but I steadied it against my leg while I, with my other hand, got hold of shrub and thereby saved myself a descent of some 500 feet. Procured quite a lot of pitch. Our descent was anything but pleasant, with sliding and climbing we at last reached the bottom. Returned to camp just as the boys sat down to supper.

The walls of the cañon range from 900 to 3000 feet, sloping backward, very ragged and craggy. The wear and tare of ages have formed innumerable ridges which run at regular intervals from top to bottom. The gulch between two ridges resembles the Devil's Staircase. The mountains here are covered with red pine with a mixture of pinions.

August 13

In camp, being Sunday. Prof., Steward and Jones climbed out for observations. Returned about one o'clock. Rest of the day spent in resting.

August 14

Started 7:30. Immediately after starting had to get out in a rapid and guide boats through rocks. Little farther down ran through a narrow channel. The *Nell* let down by line while Fred and myself held on to the side of the *Dean* and got through all right. Ran three more when we came to a very bad one, the water fairly boiling over the rocks. Concluded to make line portage, Fastened a line to her stern and bow to hold her with while two men guided her through the rocks. Being very deep in some places they had to hold on to the sides and float along with her. Got them over all right. Had dinner on the right bank under a cotton wood tree. Beaman took pictures of walls, which Prof. called "Fretting Water Falls." Fell some six feet. After dinner ran a bad rapid of ¼ mile long. It fell some ten feet. In one place could fairly feel the boat sink down. This was the commencement of the rapid. It fell abruptly three feet. It was swift riding all the way, the boats bounding like a deer. Got through all right. The *Nell* and *Cañonita* both struck rock but not damaged. Landed at the head of another. Let boats down by line. Camped on the left bank under some cotton wood. Just before getting into camp had to get out and push boat over a riffle. Ran five and seven eights of a mile.

August 15

Started out 10 AM. The reason for not starting sooner was on account of Beaman taking pictures. Ran down a quarter of a mile when we came to a fall, Called it Five Point Fall, on account of five mountains bring[ing] their ridges together here. Let boats down by line one at a time. I barked my shin very bad while guiding the *Dean* through. While letting the *Nell* down the Deacon and myself had hold of the rope when, in the middle, he slipped and fell, pulling me in with him. The consequences was a ducking. We looked like two wet bulb thermometers. Got over all right when we came to another which we had to run, but struck a rock. Fred and myself got out and guided her most the way ¼ mile long. Got in, ran half mile farther, where we camped at the head of another rapid for dinner. Started out immediately after by letting the boats down by line. Got out in about an hour and a half. Pulled for half a mile, when we came to another fall. Let down by line. Took us rest of afternoon. Camped on the right bank. Made 3 ⅜ miles. Mountains very irregular, being all cut up with lateral cañon, the reason for so many rapids, the debris being washed and tumbled into the river.

August 16

Remained in camp. Caulked boat and pitched her seams. Prof. and Jones went across the river in order to climb a mountain but returned in about an hour after starting, not being able to see far enough on account of the hazie atmosphere.

August 17

Prof., Jones and Steward climbed out for observation, while Fred and myself climbed out for pine gum, of which we found some six pounds mostly from pinions. Returned about 10:30, while the others did not get back till about 1:30 PM. After they had eaten their dinner we started out. Had not gone over a half mile when we struck a rapid which we ran—the *Nell* and the *Cañonita* ran over all right while the *Dean* struck and hung on. Fred and myself got out and shoved her over. About one-quarter mile farther down we came to a fall of some seven feet. Let down by line while four men guided her through the rocks, sometimes walking on rock while others had to hang on to the boat for dear life. Made the portage in thirty minutes. The party two years ago ran this but got swamped and lost their barometers and three rifles. The former were recovered by a long search. Pulled out, but could hear the roar of another fall ahead, which we found about a half mile farther down. Landed and let down by line. We had hard work getting the boats over. Water very swift and fall some eight feet. Got over all right. About a quarter of mile farther we had a rapid which we ran but got hung. Got out and pushed boats over, got in to pull. Landed at the mouth of Nine Mile Creek just at the head of another bad rapid. I walked up the creek for some ways. It is a nice little stream of pure water, started by springs. Who ever gave it that name I do not know. We let down by line. The *Nell* tried it without line but got hung when the current caught her bow and swung her round like a top, striking on a rock at the same time, which nearly turned her over. A little farther down she struck heavy starting her planking. Camped on the right bank. Opposite our camp is the most highest vertical cliff we have seen on the river. It is some 2800 feet high. On the top is a very singular rock. It resembles the cabins of this country. We called it [Log] Cabin Cliff. Mr. Beaman photographed it. The mountains through which we are passing at present are not laid down on the map as such, but as a platteaux, but by old hunters called the Little Mountains.

August 18

Mr. Beaman, Clem and myself climbed a small mountain for the purpose of getting a view of Nine Mile Creek. Took four pictures and started back for camp. Got in about 10. Packed our traps in the boat and started out. A quarter of a mile from camp we ran a rapid all ok. A quarter farther down we ran a bad rapid of ⅛ [mile] in length. Water very swift. Fell some 8 feet. Landed at the head of another to see where it could run. Found an appearing [illegible] channel, but the *Dean*, in running over got it a little too far to the left, and the consequence was that we struck a rock. Fred and myself jumped out on the rocks and shoved her over. No damage. Little farther down we heard the roar of a fall. Landed at the head. Took cooking utencils and while Andy prepared dinner we let the boats down by line. All safe. Came back, had dinner, carryed the cook's thing down and pulled out. Noon pulled out and ran a bad rapid. Little farther still another. Landed at the head of a fall, let down by line. Called it Melon Falls. Water fell some nine feet. Pulled out for half mile. Stopped to examine a rapid, which we ran. Full of rocks. The *Nell* struck but no damage. Landed at the foot of a rapid on left bank where we camped for the night. Made ⅝ mile.

August 19

Pulled out this morning with rapid in sight, which we ran. In going over the *Dean* struck but slid off all right. Landed at the head of another. Found it an easy task to run over if we struck it in the right place. Happily, we did and went whooping for ¼ of a mile. ⅛ farther we landed at the head of a fall, let down by line while the *Cañonita* came over. She got away from Andy, that is, she swung her bow round. The current took her round in a jiffy, pulled Andy to the Ground but sustained no injury. Pulled out for half mile to the head of another rapid. Concluded to run it, which we did in good order. Half mile farther down landed on the left bank at the head of a rapid, went over to the right bank and let down by line. Got over all right. Pulled out and ran for a mile and a half in smooth water, when we landed on the right bank at the head of a fall. Let down by line, pulled out, and landed at the head of an old stunner. Had dinner and then let down the *Cañonita* for Beaman to take pictures of the boats, as they came over the falls. Then came the *Nell* and the *Dean* last. At this fall a brook comes in on the left. Clear, sparkling water comes down over the rocks, boiling. It is started by some springs. Called it Chandler Fall and Chandler Creek. Fall of water 12 feet. Pulled out leaving the *Cañonita* behind for Beaman to fix up his pictures. Ran but a little ways when we landed on the left bank. The *Cañonita* soon hove in sight but landed to get some views of the brook, while the *Nell* and the *Dean* pulled out for a short distance. Landed on the right bank at the head of a rapid, where we camped for the night. Prof., Steward and Jones climbed out, returned at dark. The *Cañonita* came in a little before.

August 20

Sunday. Pulled out this morning, not being able to get a rest here on account of the many ants. Crossed over and let down the rapid by line. At the foot was still another, which we concluded to run. Got over all right, though it was chance work. *Cañonita* stayed behind for Beaman to get a photograph of a natural bridge which spanned a gulch some 200 feet wide and 300 ft. deep, about 1500 ft. from the river. This bridge has been formed by erosion. Camped on the right bank under cotton woods. Traces of Indian encampment could be found in the shape of wickiup poles and brands of wood. I spent the day fixing moccasins. 1 ¼ mile.

August 21

Pulled out this morning for half mile when we came to a rapid. Ran in and jumped out in the centre, where we struck. Pushed her over and landed at the head of a fall. The *Cañonita* ran over all right. Let down by line and pulled out to run another. Got in but little ways when we had to get out and push boat over, but soon after struck a channel. Jumped in rode at the rate of 20 miles an hour for an ⅛, went on for ⅜(?) more in quiet water. Landed at the head of a rapid on the left bank. Walls of the cañon running down. Distance between walls about ¾ mile. I think by the appearance we will soon be out of this cañon of Desolation. Let down by line, jumped in and pulled across to the right bank. Let down another rapid by line. Pulled out, ran across a riffle. Struck in, going over, landed at the head of another rapid on the right. Boats let down by line, then got in and pulled for 3/4 mile in smooth water. Landed on the left bank at the head of a rapid. Let down by line. Camped on the left at the foot of the rapid for dinner. Water fall some six feet in ¼ mile. After dinner pulled out for ¼ mile, landed at the right bank. Prof. climbed out while we waited for him. It rained, felt chilly. The first rain we have had for three weeks. Rain continued only half an hour. Beaman took some pictures of a vertical cliff. Waited for him two hours, then pulled out and ran three rapids, two bad ones. Landed at the head of a fall, let down by line. Fall of water six feet. Pulled out for half mile, landed on the left bank to examine another rapid. Pulled over to the right and camped for the night, under some cotton wood trees. 6 ¼ miles.

August 22

Pulled out this morning. Commenced with running a rapid half mile long. Made it in just two minutes. It was in the shape of a third of a circle. Landed on the left at the head of a fall. Let down by line, four men guiding the boats through the rocks. Pulled out through another rapid, landed at the foot to let down over another. Pulled out for ¾ mile Landed at the right bank at the head of a rapid. Ran through it partly, then landed and let down the rest of it by line. Pulled out for a mile and a quarter in smooth water. Landed on the left at the head of a fall. Let down by line, four men guiding boats. Passed rocks, camped for dinner at the foot of a fall on the left. Pulled out at 3. Ran a huge rapid. In going over the *Dean* struck twice. Jumped out and pushed her over. No damage. The water, in rushing through a narrow shoot, caused the waves to boil up some six feet. Ran another about three quarters mile farther down, then landed on the right bank. The country all of a sudden opened out on the right. Saw an Indian pony turned out to die. No Indians being in sight, and the horse being lame, causes us to conclude that he had been left. At first we thought that the Major might have come down here, but the manure seemed to be very old. Pulled out, ran a rapid, landed at the root to examine another, which we concluded to run. Pulled through and landed on the right bank at the head of a fall. Camped for the night.

August 23

Prof. and Jones climbed out and returned at dinner. Took in a feed of sowbelly in our stomachs and loaded the boats with our baggage. Let boats down by line over the falls. Called them Sharp Mountain Falls. Beaman photographed an over hanging cliff of stone 1200 feet in height. Pulled out for a mile in smooth but swift water. Ran a rapid half a mile long, landed at the head of another. River split up in two channels. The *Nell* took the left but ran aground just ahead of the rapids. Prof. told Jones to take the right channel, which we did. An old tree had fallen off from the bank and lodged. The current passed through this tree, and not being aware of the great suction, was carried into the tree, struck on Fred['s] oar lock, which it snapped like a pipestem, and there hung under the tree. I removed my oar and lock then pushed clear and we went at the rate of 30 miles an hour for a quarter. Landed on the left bank ¾ mile farther down and camped for the night. While cutting some willows for my bed I discovered a huge rattle snake coiled up on the roots of willows. I called to some of the boys to bring a pistol, when Steward brought the Prof's, and shot him through the side of his belly, but a second ball passed through his head. I threw him on the beach. Counted nine rattles and a button, which made him nearly ten years old.

August 24

Pulled out this morning in smooth water for ¾ mile. Ran a rapid and then ran in quiet water for nearly two miles. Landed at the head of a fall on the left but pulled to the right and let down by line. Got an awful rap on my shin. Lost a handkerchief and bandage which was around my leg. Fall of water 10 feet. A nice little creek coming in just above the falls on the right started by some springs some 10 miles above its mouth. Pulled out, ran the rapids then had a quiet river for nearly 3 miles. Landed at the head of a rapid on the left. Shot at some otter but failed to get any, being the first we had seen on the trip. Let down by line. Water fall about 7 feet. Pulled out into another rapid but failed to make a clean run. Struck as we went in, jumped out, pushed over and then pulled through for ¼ mile. Ran another a little below, and still another about 100 yards below. Made them all OK. Landed on the right for dinner, pulled out at 2:15 and ran a rapid immediately after starting. Went whooping for a mile, then another rapid. Ran it then for a mile and a half in quiet water. Landed on the right at the head of a fall. Let down by line for little more than one quarter, pulled out for ½ mile, then ran a rapid. Pulled out and landed on the left at the head of a rapid. We have seen coal all day. Concluded to be out of the Cañon of Desolation, and are now in the Coal (Lignight) Cañon.[20] The features of this are the same as in Desolation. It is God's country, for man don't want it. The walls in this cañon are narrowing up again. The walls are perpendicular but not high. Ran the rapid which proved to be an old hummer. Made it all right. Ran another, then had shallow water for ¾ mile, then ran another rapid. Landed on the right to examine one which we ran. Struck as we went over. Landed on the right. Camped for the night.

August 25

Prof., Jones, and Capt. climbed out, returning at 11:30. Reported that we were running out of cañon. Had dinner and pulled out. Ran two rapids immediately after leaving camp. Landed on the right at the head of a fall. Let down over two falls. First fell about six feet and the second five. Ran along for a mile, then came to another rapid. Ran it and four others all OK. Came to the mouth of the Little White [Price River], about 60 ft. wide, but at this season of the year contains no water. We placed our beds on a shelf of the river bank. The Valley of the White is rich in scenery. It is called Castle Valley. The mountains on each side are so much eroded that they form all sorts of castles and fortresses, hence the name. Immediately on the bank of the river grow box elder, cotton wood, willows and grass.

August 26

Pulled out this morning, started with a rapid half a mile long. Ran it all right. We shot over it like an express train. Pulled a half mile in smooth water for a mile and a half. Landed on the right to examine a rapid. Ran it little farther down. We found two more close together, two old hummers. Ran them like a demon. The *Nell* and *Cañonita* shipped half full, but the old *Dean*, happy-go-lucky, went over without any. Little farther down we ran another, landed on the right to examine a rapid. Proved to be an old buster, but ran her all right. Ran along for two miles, then ran another, Little below, we landed on the left to examine another. From this point we can see the end of Coal Cañon. Ran it all right. Just below the rapids Capt. shot a beaver. Had hold of him twice, but each time slipped out of his hand, when he finally sunk. Little farther down had a rapid which proved hard work getting the boats over. The river at this point covers nearly one quarter mile in width. Worked nearly an hour. Ran little farther, then camped on left bank just inside the cañon. Have to wait for the Major till the 3rd of Sept. After dinner hauled up our boats to let them dry for caulking.

August 27

Sunday. Dryed rations and fixed my clothes. Fred and myself went down to plant our flag at the foot of an adjacent island as a signal for the Major.

August 28

Caulked and pitched boats. Caught three large fish, one weighing 15 pounds. Had him for supper.

August 29

Commenced fixing my moccasins. At dinner Beaman, Fred, Clem, Hatten and myself were discussing the Major's probable return, when all of a sudden three shots were heard—his signal. Everybody was up in a moment, answered the signal, and in a jiffy Prof., Bish and myself were across the river with a boat. A few minutes after we saw the Major and Mr. Fred Ham[b]lin on the brow of the hill. Soon pulled them over to our camp. Major gave orders to pack up and start down some five miles where he had left the pack train. Soon were skimming over the water, while Jones went round by land with the horses. Ran three rapids. The *Nell*, in going over, got foul of a ledge of rocks but sustained no damage. Arrived at Gunnison's Crossing, so named after Captain Gunnison, who in '53 crossed the river at this point to explore the country around here, and farther west in Utah he was killed by Gosi Ute Indians at Severe [Sevier] Lake, for a long time called Gunnison's Lake. He had strayed a little ways from his party when they killed him. Found two letters for me from my brother in Frisco.

August 30

Spent the day writing and tinkering.

August 31

Major, Beaman, Clem and myself went out with the instruments to take a picture of Gunnison's Butte. Returned at dinner and immediately went out again to take more views. Returned at 5 PM.

September 1

Mr. Ham[b]lin and his nephew made preparations for returning to the settlements. At noon they left, taking with them our mail, fossils, and other loose baggage. Soon after we dropped down stream half mile, being camped on the right. Moved over to the left in order to be able to get back to our last camp in the cañon where the *Nell* had left a saw. Beaman and Fred went after it. Returned at dark. Camped for the night.

September 2

Pulled out at 8 AM. Ran along through Gunnison's Valley. Came to many shoals. Jumped out very often to put boat over, but no rapids. Camped on the right bank for dinner. Immediately after pulled out and stopped for barometrical observations. At the Spanish Crossing at 2 PM. Someone had crossed here lately, as fresh horse tracks could be seen on the gravel, also a barefooted boy's print in the sand. Pulled out soon after observation. This trail formerly was used quite extensively by the Spaniards in going from Santa Fe to Los Angeles, Calif. Ran down some four miles and camped on the left side in a cotton wood grove.

September 3

In camp. Major and Jones went across and went to some bluffs about 9 miles from here. Returned at night. Major brought a lot of fossils. Steward and Bish. went out some four miles. They found a cave of some two hundred feet in length. Made myself a pair of moccasins.

September 4

Pulled out at 9 AM. Still running through what the Major calls Gunnison's Valley, but now and then would pass through a line of bluffs, highly colored. Made ten miles then camped for dinner on the right bank under a sandstone bluff. Country is barren and lifeless. Pulled out again at 2 PM. Saw a beautiful little butte in the shape of a cathedral. Beaman took a photograph of it. Major called it "Dellenbaugh's Butte" after Fred. Ran down some 5 miles when we came to the mouth of the San Rafael, a stream of about the size of the White. Starts at the Wasatch Mountains. Camped on the right hand.

September 5

Major and Jones started up the San Rafael to be gone two days. Boys found a lot of arrow heads. This part of the country is not much traveled at this season of the year as the game is all upon the mountains, but in the winter the valley is a good place for game.

September 6

Major and Jones returned at 10 AM. Bish and Clem started out at 7 AM for a butte some ten miles from here on the left bank. The peaks which the Major and Jones visited are called "Jones Peaks." This part of the country is not laid down on the map and is left blank, but on our return will be filled up. Helped Jones to measure the distance from mouth of San Rafael to Jones Peaks. Found them to be ten miles. At this place the Indians make their arrow heads as we found their tools, stones, and a lot of chips. Bish and Clem did not return. Kept a fire burning all night as a signal for them, thinking that they had lost their trail.

September 7

Major and myself started out in search of Bish. and Clem. Found them on their way in some five miles from camp. Struck for the river where Prof. had come down with the boat. It began to rain fearfully when we reached the river, and a little while after hundreds of streams came shooting from the top of the bluffs. One in particular came shooting over a vertical wall 200 feet high. The spout was some eight feet in diameter. It looked like a river of red mud rolling down, having come from flats of red sandstone dust. It was a grand sight. We took shelter under an overhanging wall. Had dinner and then pulled out. Still raining. All the way down streams came shooting down, some bright as cristal while others were of a reddish or umber collor. Camped on the left bank in an oak grove. First oak seen on the trip. At noon we ran into what the Major called Labirinth Cañon. Walls are increasing in height as we go down. A very peculiar feature is noticeable: the wall on one side is vertical while on the other it is rounded and sloping back. It changes at every bend of the river. In the bend the walls are vertical while on the other they slope and continue so until the river makes a turn. Stopped for half an hour, for Fred to do his sketch, being unable to do it in the rain.

September 8

Beaman took a photograph of a tripled alcove [Trin Alcove], and after climbed up on top of a bluff from which he took several pictures. In the afternoon crossed over and took several more.

September 9

Pulled out about 9 AM leaving the *Nell* and the *Cañonita* behind. Landed about two miles from camp for the Major to geologise. Pulled out after stopping half an hour. Ran several times aground. About noon it began to rain. Landed on the left. Pitched our paulin, built a fire, and waited the coming of the other boats, who soon made their appearance. Had dinner and pulled out. Walls increasing in height as we go down. About 4 o'clock the *Dean* got into a channel which proved hard work for us. Ran aground, signalled the other boats to take the left hand channel. All the *Dean*'s crew got out to push her over. Dragged her for some twenty-five feet but being unable to get her any farther we had to call on all hands. At the end of the bar we had only about three inches of water, so we had to lift her over some twenty yards more. Ran until 6 PM. Camped on a sand bank in bow knot. Made 16 miles. The river at this point bent back and forward forming a bow knot. Major called it Bow Knot Bend.

September 10

Beaman climbed up on top of a ridge about 300 feet high and a thousand wide, and four miles long, round which the river doubles, forming the last loop of the bow knot. Pulled out at 10:30 leaving Beaman on the ridge photographing. To meet us on the other side. Got on the other side a little before 12 Noon. Camped on the left for dinner. Concluded to stay for Beaman to fix his chemicals. Major and myself climbed out on the ridge while Prof. Jones, and Steward climbed out on the left.

September 11

Pulled out at 8 AM. Ran down some six miles and camped on the right bank for dinner. The walls of the cañon from 800 to 1000 feet in height. Very much cut up, forming numerous alcoves, pinnacles, columns, peaks, towers, castles and cliffs. Water placid and very shallow. Pulled out at 3 PM. While preparing dinner the boys dried out their blankets which had got wet during the last night's dew. Stopped on the right for Beaman to take a picture of a latteral cañon. While some of us climbed up, the *Nell* pulled out to a place where the cañon breaks down some five miles from here. As soon as Beaman had taken his view we started after and overtook them just as they landed to camp, being six o'clock and getting dark. Camped on the left.

September 12

Pulled out at seven o'clock. Ran down some seven miles. The walls of the cañon are broken down very much and I think can hardly be called a cañon, being in some places only about 200 feet, while on the other points it is level with the river excepting a bank. Saw a curiously eroded mountain or rather a cliff. It was in the shape of a Greek Cross [Butte of the Cross]. Beaman photographed it on the right bank. Major, Prof., Steward, and Bishop climbed out on the left while we dropped down some 400 yards to prepare dinner [near Barrie Creek], leaving Beaman to come round the point by land. The whole party united at one PM. Had dinner. The Major, Beaman, Jones, Fred and Clem took the *Dean* downstream about a mile, while we put up our cooking utensils. Soon followed Andy and myself taking down the *Cañonita*. Major came down soon. Started down with the *Dean*. Prof. came down with us some three miles and went into camp on the right bank. At this place the river has cut through a wall of sandstone round which the river formerly flowed, making a cut off of some four miles. The bend [Bonita Bend] on each side was to the side of the wall. The old riverbed is distinctly visible. Left our axe at dinner station. Major concluded to call this the end of Labyrinth Cañon. The break in the walls extends over some nine or ten miles.

September 13

Major and Prof. climbed out, while Beaman took some views. Pulled out at 11:45. Ran four miles and a half. Camped for dinner on the right under a shelving of rock. While eating dinner heard the bark of a wolf a little ways down in the bend of the river. Some of the boys mistook it for the bark of a dog, thinking it might be from some Indian camp. This part of the cañon is very wide, forming quite a little valley on each side. Walls are overhanging—that is, the homogenus sandstone projecting over the red sandstone which came up at last night's camp. After dinner pulled out for a mile and a half, when the Major and Prof. climbed out. Beaman took a picture of a valley of rocks. I never saw anything to equal it. It was nothing but peaks, pinnacles, towers, spires and chimneys. Pulled out again at 4:30. Ran down about seven miles, then camped on a high bank. The cañon since four o'clock is narrowing up—perpendicular and very ragged. Height 800 to 1000 feet. The country back is cut up, forming numerous terraces, gulches, alcoves and winding stairways.

September 14

Major and Steward hunted for fossils. Beaman took a picture of the cañon, from a bend a little below camp. Major, Prof. and myself pulled up stream half of a mile and obtained a lot of fossils. Returned at noon. After dinner pulled out but ran but a little ways when the Major got into a fossil fever. Stopped for 45 minutes. I found a large fossil tooth which I gave to the Major. Pulled out till four PM. Landed on the right. Boys found a lot of Moqui ruins, perhaps, some 200 years old. This ancient race has been driven from country to country on account of their industry—cultivating the land and raise enough to subsist on. Make pottery, blankets and everything that they require. The other tribes don't like them because they work.

September 15

The *Nell* pulled out for the junction of the Grand and Green for Prof. to get fornoon's observations. Major, Jones, Hatten and myself climbed out this morning. While going up a gulch Hatten found an Earthen pot full of split willows and reeds, the work of Moqui Indians. Had a hard time getting out onto a ledge of rock on which we had to get out in order to reach the top. At that place the Indians who resided here used to climb out, for two cedar trees were set against the wall looking old and decayed. By their assistance, I should judge when they were green, could reach the ledge. By the assistance of some rocks which were also piled by the Indians, we commenced our ascent. Jones, being the tallest, got on my shoulder, and so reached the shelf. Next came the Major, whom Jones pulled up by a rope brought for the occasion, then Hatten. Myself, having moccasins on, I shimmied up like a lizard. After reaching the top walked out two miles to a little butte, where we took bearings, and returned at noon. At the shelf we lowered ourselves down with the rope. I was the last. Lowered myself as far as I could when Jones pressed my feet against the wall and slid down. After dinner pulled out, reached the junction at 4 PM. The Grand is a stream about the size of the Green and at present has more water. United they form the Colorado. At the junction a Mr. —— of Colorado is supposed to have made a treaty with the Indians for the possession of this valley, with a site for a city at the junction. The government furnished Mr. —— with necessary funds to carry out the treaty, which was supposed to have been signed by the great chief, Black Bear, and a few other prominent chiefs of tribes, under a cotton wood tree. But where the valley, the site for a city, or even the cottonwood trees, I can't see. The walls at each angle come close to the rivers, leaving a narrow bank of drift sand on which a few willow trees seem to flourish. Height of walls 1300 feet, nearly vertical. At each angle the strata is turned up making the dip nearly 25 degrees inward. Major concluded to call the last cañon from the break in Labyrinth Still Water Cañon.

September 16

Major, Jones, Beaman, Fred, Clem and myself pulled up the Green some ¾ mile and climbed out. Beaman took a lot of pictures. Fred Sketched. Clem and myself gathered a lot of pitch which found to be plentiful. From this point we could see the Sierra La Salle stretching northward. A grand view is presented to the eye from this point, of craggs, buttes, spires, and pinnacles in various shapes and forms and color. Major and Jones returned at 3 PM. Reported as having seen parks formed or enclosed by pinnacles. Returned to camp, leaving Beaman's apparatus on top. Got two gallons of pitch.

September 17

All hands excepting Bish and Jones started up for to see the parks. After an hour's climb we reached the top then walked two miles to the parks. Such a sight I shall never forget. I counted five parks enclosed by pinnacles formed by erosion. They looked to me like monuments in a cemetery. Everything looked sombre and death-like. Nothing disturbed the scene except the sighing of the wind or the falling of a chip from the rock. The water collects in a large basin in the center of each park and from hence into a gulch into the river.

September 18

In camp. Major and Jones climbed out on the left angle of Colorado and Grand. Returned late at night. A breach of some 50 feet wide was visible on the right of the junction, but the current cut it away in three days.

September 19

Pulled out this morning at 10. Divided all the rations before starting. Ran down some four and a half miles when we came to a huge old rapid. Landed on the left and let down by line, each crew working their own boats, Fred and myself wading by the side of the *Dean* while Jones held onto the rope from the junction. This cañon is called "Cataract", and well it deserves its name. The water, while going over a fall or rapid, it goes down like steps of stairs. Let down one more before dinner. Immediately after dinner let down another half mile farther—another old buster—let down for half mile—then ran a half to the head of a fall, let down, got in, pulled for

¾ mile when we ran one picking our way through rocks. Also ran two small ones. Came to a fall, let down and picked through the rocks. Got in and pulled through two almost joining. Went into camp on the left.

Major and Fred climbed out on the right. I took them across at the head of the fall. Bishop and Steward went out on the left for topography and geology. Prof., Jones, Clem and myself let the *Nell* down by line on the right side, Prof. and Jones at the rope, Clem and myself guiding her through the rocks. About half mile down we came to some rocks which projected out into the river from shore, immediately below an eddy. Being afraid that she would run up under the fall and fill, so we attached the other end of our rope to the stern for Prof. to pull her in short, while Jones, Clem and myself held onto the bow. Down she went. Prof. pulled her stern in a little too much, which swung her bow out, and in a jiffy the current caught her bow, and no ten men could have held her, the rope being too short, or else we would have held her. After turning we had to let go or be dragged over the rapid. We chose the former. Soon the rope tightened in Prof's hold. He being unable to resist the pressure, let go too, and the *Nell* went fluking over the fall, while we kept up the race with her on shore over the rocks. She went over all right. Immediately below two more falls were visible. Fearing that she would go over these, we ran for dear life—visions of short rations could be plainly seen. Prof. started in but backed out. I plunged in and swam to her almost breathless. I gained her deck twenty feet from the head of the lower fall. I reached an oar to Clem whom I pulled in. A counter current took her far enough for me to reach an oar to Prof. who pulled us in just in time for to save us going over the falls below. Went back and let the other two down on the left all right. Camped on the right on a sand beach. Major and Fred returned about 5 PM.

Pulled out at 8. Good river for ¾ mile then came to a rapid. Ran it and five others, then let down three old hummers, four men guiding her through the rocks. Ran two more, then camped on the left at the head of a fall for dinner. About 1:30 we let down by line. Walls of cañon from 1000 to 1700 feet high, very ragged, composed of limestone, and are still increasing in height. About a mile farther down we had another let down. Wading by the side of the boat, sometimes wading while at others hang[ing] on for dear life. Took us nearly two hours. Immediately below another. Major, seeing how wet, cold and tired we were, concluded to run it at a great risk. Pulled out into it, but no sooner in than we struck a rock, nearly keeled over, but fortunately slid off. Pulled for half mile through foaming cataracts. Shipped our standing room nearly half full of water. Camped on the right on a sand beach.

September 22

Commenced by running a small rapid then let down two. At the first, had to get the boats out of water and slide them over rocks. Broke our keel plating. Worked all forenoon to make two portages. How Mr. White[21] came through this cañon on a raft in one day is more than I can, but such is laid down in history, but I don't be-liefe it. A raft could not live—it would go to pieces while going over the first cata-ract. Dinner-camp on the left, under a hackberry tree. Walls opposite vertical and 200 ft. high. Let down another cataract. Took us until 4 PM. Went into camp on the right—sandy beach. Repaired the *Nell* and *Dean*.

September 23

Let down over an old snorter. Ran a half mile, let down another. Got in and pulled through one all right. Stopped at the head of a rapid. Took out the cooking kit and let down while Andy prepared dinner. I took the Major down in the *Dean* to head of a fall, tied up and returned to camp where we had to wait for the boys, having left them below the first fall, for Beaman to take pictures. Soon the *Cañonita* came in. Had dinner on boat. Dropped down to the head of a fall. Major concluded to run it, which we did in fine style. Beaman took an instantaneous view of us as we went over. Fall of water, 12 ft. in 50 yards. Pulled to the head of the next fall, where we waited for the other boats. Had a mile of swift water. Walls very craggy and verti-cal on top. The other two boats soon came. Crossed over to the left and let down by line for half mile. Hard work getting boats over. Took us the rest of the after-noon. Fall 70 ft. in half mile. Camped on the left at the foot of fall.

September 24

Pulled out this morning into a rapid, made it all OK, but to be stopped by a fall. Let down by line, stationed all hands but two along the fall below, with the rope to pull her in when she came over. Two men shoved her off. It put one in mind of a bucking mule to see the boats jump through the waters. Got in and pulled out for a mile and a half through rapids. Done some wild riding through the rocks but got through all OK. Stopped at the head of a long rapid, let down for half mile. In going round a big rock the stern of the *Dean* caught on a sunken rock and nearly capsized and filling her half full of water. With the assistance of two men, slid her off, got in and pulled through a rapid half mile long, made it all OK. Little farther down ran another, then came to one where we let down. Being full of rocks, got in and pulled through another. Landed at the head of a cataract. Let down by line just below. Went into camp on the left, hauling our boats up to dry. Had dinner, then fixed around my clothes and shoes. Major, Jones and Beaman had a walk up a lat-eral canyon [Gypsum Canyon]. Returned late at night.

September 25

Major and Fred climbed out for observations, while Beaman, Jones and Clem, went up the cañon for pictures. Bishop working at his map. Prof. and Steward pitched the *Nell* and then went up the side cañon. I pitched the *Dean* and helped Andy fix the *Cañonita*. All but the Major and Fred returned at 5 PM.

September 26

Pulled out this morning to the head of a cataract little below camp. Let down by line. Got in and pulled through into rapids, ran a little ways, then landed on the right to examine another. Walls craggy and slanting back, towering up some 3000 feet. Pulled out into the rapid, and went fluking for half mile. The *Nell* while coming over, filled but managed to reach the shore in safety. The *Cañonita* struck and half filled, but sustained no damage. Little below, about quarter of a mile farther down pulled through a rapid about one-third of the way, then let down another third, then pulled through the rest to the head of a fall. Stopped for examination. Ran it all OK. Shipped some water. Fall of water 7 feet. Pulled out and ran two small rapids. The cañon is very narrow, walls perpendicular, from two to three thousand feet. Broken up in some places, forming towers and pinnacles, some of them nearly 1500 feet. Limestone at the bottom and sand on top. The former causes so many rapids, being the hardest. Landed on the left for dinner. Immediately after dinner pulled out into a rapid. Made it all right. Waited on the left for the *Cañonita*, then ran another small rapid. Landed on the right and went into camp. Beaman took a view of the cañon.

September 27

Went up a gulch [Dark Cañon] near camp for Beaman to take pictures. Returned at 10. Let down over the rapid at camp. Immediately after ran down for a mile and had dinner. Pulled out after dinner for a half mile, then came to a cataract. Let down by line, got in and pulled through a rapid just below the cataract. Ran another and then had a beautiful river for nearly four miles, then came to a cataract. Walls dropping a little in height. Ran it all right. Had three miles more of good river with three rapids in it at equal intervals, then came to a cataract. Let down by line and went into camp just below. Made nine miles.

September 28

Pulled out this morning leaving the *Cañonita* for Beaman to take pictures. Just below camp ran two rapids. Below the second we left the *Nell* to signal Beaman how to run. Pulled for a mile in nice river, then came to a huge cataract. Waited for the other boats who soon made their appearance. Major and Prof. climbed out but failed to get out. Returned about 4:30. Let down by line. After supper pulled out, ran a rapid just below camp, about a quarter of a mile farther down. Landed and let down through the head of another. Then had to sink or swim, for no footholds be had on either side. The current [illegible] is very strong toward the left wall and came near smashing us to atoms, but saved ourselves by a few feet. The other boats did not get quite so close, being aware of the strength of the current. Pulled out and ran two more. Darkness came on and we went into camp at the head of a rapid on the right bank.

September 29

Had a terrible night of it. Wind blowing the drift sand like snow, and everybody had red eyes this morning. Rained hard this morning. Delayed us for an hour, then pulled out into the rapids. Made it all OK. Raining all forenoon off and on. Ran down about five miles to Mille Crag Bend. Major and Prof. climbed out. This is the end of Cataract Cañon, and thank God, [for] it has been hard work ever since we left the junction. The walls break down and are only about 1800 feet high, all cut up into pinnacles and towers, crags innumerable, hence the name. Walls are sandstone. Camped for tonight on the left. Ran the limestone under and now [f]ace the Homo G[eneous?] at the top and bottom.

September 30

Pulled out at sunrise. Ran a rapid just below camp, another a little ways from the first. Ran about a mile. Ran two more, then ran for six miles in a beautiful straight cañon, no rapids. Walls on both sides come down to the river and are only about an eighth of a mile wide. Called it Narrow Cañon. Came to the mouth of the "Dirty Devil" River, the end of Narrow. The Dirty Devil is a stream of about fifty feet wide. Water very muddy. It starts from a range of volcanoes [Henry Mountains] about forty miles from here. Not laid down on the map. Major, Prof and Jones climbed out. Returned at night. The walls break down to nothing, running the sandstone under.

October 1

Cached the *Cañonita* until next spring, having no material for photographing and are short of provisions. Major intends to explore this part of the country, bring in material and ride down in her on the Colorado to the Par Weep [Paria River]. Hauled her up into a cave about 20 feet up from the river and covered her up with sand. Major and Jones climbed out to find a trail leading in. The *Nell* dropped down about a mile and a half for them to climb out from there. About 4 PM Jones and Major returned and we dropped down to where the *Nell* had landed. Went into camp.

October 2

Pulled out early this morning. Ran two rapids just below camp. Struck in the second, had to get out and push boat off. Water very cold. Wet clothes all day. Running up the red sandstone. Major calls them the "Orange cliffs." Landed on the left to examine some old Moqui ruins. Found one house about 20 ft. long, and the walls, what yet remained, were 14 feet high.[22] Found a lot of hieroglyphics on a smooth sandstone. Could make out a good many things out of them but none perfect or recognizable. Had dinner and pulled out for 11 miles. Camped on the left where we found another old ruin.[23] Had seven rapids, ran them all OK. Called this Mound Cañon.[24] The top of the plateau is curiously eroded. Look like huge mounds closely joined, like a grave yard.

October 3   Pulled out at 8 AM. Ran till noon. Made 14 ⅜ miles. Had a fine run, only a few rapids. Pulled out at 2 PM, and ran 13 ⅜ miles. Camped on the left. Cañon very near a mile wide. Walls running up and down from 100 to 700 feet in height. Looking from the top it is a vast stretch of sandstone eroded, forming huge mounds closely packed together, like a grave yard. No vegetation can be seen, nothing but the bare rock. What the Indians would call [Unkar] to weep,[25] or Red stone Land. Huge water pockets can be found all over, varying from two to thirty feet in depth. In one we found a cotton wood tree. It was a well half full of sand or nearly so, judging from the appearance.

October 4   Pulled out about 8 AM. Ran for two miles on a beautiful river, then came to shoals, or rather we came into a bed of hetrogeneous sandstone. The bed was as smooth as a barn floor. Had to slide the boat for nearly two miles, then changed directions and ran it up. In coming over a small rapid the *Nell* stove a hole in her centre cabin. About noon hauled her out and repaired her and had dinner. Pulled out 2:25 PM. Ran nearly 20 miles. Camped on the right under a ledge of rock.

October 5   Pulled out this morning 7:20. Ran five miles and a half. Landed on the right at the mouth of a small creek. Tasted strongly of alkali. Pulled out again after delaying half an hour. Passed the San Juan about noon. Stopped two miles below the mouth. The San Juan starts from the San Juan Mountains on the western slope of the Rockies. At dinner went up into a gulch, about a quarter of a mile, when our progress was stopped by a large temple or space containing a small pool of water [Music Temple]. The water had undermined a solid rock, making a regular amphitheatre of about 500 feet long, 300 wide and 200 feet high. The walls rounded up, almost closing at the top, leaving only a space of about 20 feet for daylight. Before entering this temple a beautiful little cotton wood ornamented its portals. At the junction of the San Juan and Colorado Mound Cañon terminates.[26] Pulled out after dinner. Came in view of a volcano [Navajo Mountain] on the left. Major called it Mount Seneca Howland in memory of his old topographer. It is about 4000 feet high and about 14–15 miles long. Ran four bad rapids scattered at intervals. Cañon very wide. Off from the river numerous huge piles of rocks can be seen, towering up to 1500 feet. Look like monuments. The Major calls this Monument Cañon. All along the walls the rock has been eroded by the drippings of springs, making it sloping, and on this slope grow small oak and flowers, forming little glens. They are very numerous. Camped on the left. 25 miles.

October 6

Pulled out this morning. Had a fine river. Found some newly burnt brushes, the tracks of a horse, and white men. Concluded that it must be the Ham[b]lins looking for us, having come down from the Crossing of the Fathers, or some disappointed prospector. About half an hour before we camped for dinner we found a fox on a few rocks in the center of the river in a rapid. Boys fired at him when he took to the water and was carried down the current. Saw him no more. Camped for dinner on the right on a bench of rock. Pulled out after dinner for the Crossing of the Fathers at which point we arrived about 4 PM. Heard a gun, watched the bank for a moment, saw a little white streamer attached to a pole, and presently saw my old friend Capt. Dodds, of Uintah acquaintance. Landed and found rations, but Mr. Hamblin had gone to Ft. Defiance, Arizona, on a Mormon mission, to obtain indemnity from the Government—the Navajos having been down to Kanab and forced the people to give up their stock and everything else they wanted. They submitted rather than have war with them. The Navahoes receive an annuity from the Government. Hamblin hopes to stop some of it. Found also two prospectors having come down with Dodds from Kanab.[27] They were anxious to reach some cañon where the water ran swift. They said that they had come all along up the river from the Virgin. Found gold everywhere, averaging from 3–5 dollars a day.

October 7

In camp. Fixing up for a trip to Salt Lake with the Major.

October 8

Still [in] camp. About a half mile round the bend is the old Ute Crossing, or more largely known as the "Crossing of the Fathers." Ascalante [Escalante] crossed with 100 pri[e]sts to be distributed among the different tribes. Nothing was ever heard from them afterward. It was supposed that the Indians killed them.[28]

October 9

Still in camp. Boys writing letters.

October 10 [29]

Started out this morning with pack train. Capt. Dodds and the two miners started back with the Major and myself. Followed the trail for about 12 miles. Went into camp near a gulch. Found water for our animals—full of alkali and salt.

October 11	Started out this morning and went about 12 miles to a spring of alkali and salt. One of the miners and myself started up the mountain for some game but found none.
October 12	Started out early. Reached the Pa Weep [Paria River] about 5 PM. having come about 25 miles. The river is narrow and very swift Good tasting water. Starts from Wasatch Mountains. Wind blew cold.
October 13	Major and Capt. started for Kanab, leaving us to bring the train. Started with the pack about 8 AM. Shot four ducks. Made about 16 miles. Left the trail and ran up a gulch for two miles to a spring and camped.
October 14	Duck for breakfast, and pumpkins which we had brought from the farm. At the Pa weep where the Mormons try to make a settlement [Paria] having built a few homes, walled in. Made Kanab about 6:30, having come 30 miles.
October 15	Kanab is a place of about 25 settlers, on the Kanab River. Kanab means "willow" in Indian. It sinks in the sand about six miles down the valley. Had a long ride this morning after our stock, returned at noon. In the afternoon the two miners dissolved partnership. Riley was hired by the Major to take rations down to the mouth of the Pa weep. Beaunamat [Bonnemort] intends to go to Beaver.
October 16	Fixed wagon covers and got ready for starting tomorrow. Major bought Bonemote's outfit and he goes, to Salt Lake with us.
October 17	Hitched up our two mules and sailed out Took road by way of Scum pa [Skutumpa] the Indian for it. It means Rabbit bush water. Went thirty miles to J. D. Lee's farm, who entertained us hugely.[30]
October 18	Started this morning for the head of Severe [Sevier River] which place we reached at night. 30 miles. Camped by a large spring by which this branch of the Severe is started.
October 19	Had a terrible storm last night. Our blankets wet through. Found my left side in water. Top blanket froze stiff as a poker. Started out and made Pang wich [Panguitch], 30 miles, a place of some 300 settlers—Mormons, of course. Pang wich means "fish" in Indian.

| October 20 | Had a mule show and then started out. Just after leaving town saw a flock of sage hens. Bonemote shot two fat hens. Met a lot of Indians. Major traded two pipes for his tobacco. Pipes cut out of red clay. First I have seen of the Pai Ute make. They generally get them from the Snake Indians, who are very expert at pipe making. Went through a Cañon of the Severe. Rough road. Crossed the river about ten times, but only 12 miles in length. Went to a deserted town from which the Mormons had been driven by the Indians.[31] Called Circleville. A beautiful valley. |

| October 21 | Started out. Passed through Marys Vale. Crossed the neck of a mountain. Reached Alma [Monroe] at 7 PM. Came 42 miles. Were entertained by Bishop [Moses] Gifford. Alma is situated in a beautiful valley, but every [illegible] is done by irrigation. Fine streams come from the Wasatch, clear as crystal and full of trout. |

| October 22 | Started out this morning early, and went to Glens Cove [Glenn's Cove, now Glenwood] to Bishop Neberker.[32] Got there about 10 AM. Had dinner and started out for Willow Creek [Mona]. Came 40 miles. |

| October 23 | Started early. Reached Manti by noon. Here we heard of the trouble at Salt Lake.[33] The Bishop was all in a flurry. Indians came in and told him that soldiers were on their way to arrest him. We left him in a state of boiling conditions. Started for Spring Town [Salem?] which place we reached about 5 PM. Came 28 miles. |

| October 24 | Pulled out early. Reached North Bend [Fairview] at 9 AM. Crossed the mountains to Spanish Fork, at which place we camped. Another surveying party came to camp with us. Had a pleasant time. This party, under command of a Mr. Fessem [?] intended to survey the Valley of the Severe.[34] |

| October 25 | Here Capt. Dodds left us for Uintah by the Spanish Fork Trail. Left the other party in the arms of Morpheus. Went to Springville, which we reached at 8:30. Found the stage had not come. We drove to Provo, the second largest city in Utah. Reached it at 9:30. Had come six miles. Here the Major took stage, while we fed and started for Leehigh [Lehi] where we camped. 39 miles. |

| October 26 | Started for Salt Lake, which place we reached at 2 PM, having come 35 miles. |

| October 27 | Stayed in the city 14 days, purchased 11 animals and two wagons. Brought Mrs. Thompson and Mrs. Powell back with us. Left the city on the 10. of Nov. Arrived at Johnsons Cañon on the 19th.[35] Stayed one night—our party consisted of Ma- |

jor, Capt. Dodds, Bonny [George Otis] McEntee,[36] Mrs. Powell, Mrs. Thompson, Lavina Neberker [Nebeker], a young miss who joined us at Glens Cove,[37] and myself. Mr. Thompson met us about four miles from Johnson. Next morning I went to Kanab for flour and raisings. Found Steward, Jones and Beaman. Steward returned with me. Same day moved camp about two miles. Stayed two days.

November 22 — Moved to Eight Mile Spring.

November 23 — Fixing camp. ~~Steward left for home on account of ill health.~~ [Crossed out in original.]

November 24 — Capt. Bishop, Fred, Clem, Andy and Ryley removed from House Rock.

November 25 — Still in camp hunting horses. Jones came down.

November 26 — Prof., his wife, Jones, Bishop, Clem, Andy, McEntee left for Kanab Gap. Ryley and Capt. Dodds left for Kanab Wash, but camped with Prof. who retained them until the Major went down.

Moved our camp in Dec. below the Gap. from this point a series of observations was commenced, building monuments and measuring a base line ten miles long. Steward left us on the 4th of Dec.[38] Major, Dodds, Ryley and John Stewart[39] of Kanab went with the Major to find a way down to the Colorado by way of Kanab Wash.[40] Returned in about eight days time, having been successful in getting to the River. After his return he concluded to go East so of course I got everything in readiness.[41] Discharged Beaman. Major and him had a falling out.[42]

1872

February 2 — On the 2nd of February with four in hand drove Major and his wife and Lavina to Beaver. Camped at Pipe Spring with Bishop [Anson P.] Winsor.

February 3 — Had an awful time finding horses—did not get started till about 12 AM. Camped five miles west of Short Creek.

February 4 — Got off early and drove to Toquerville, stopped with Mr. Neberker.

February 5 — Snow and hail this morning, but nevertheless hitched up and drove to Kannarah [Kannaraville] 25 miles. I nearly froze. Stopped at the Bishops, but he not being at home nor any of the male portion the young ladies of the house had gone to a dance, nobody but the old lady being at home. Major, his wife and Lavina occupying a room of the parlour while I stretched my weary limbs on the floor of the parlor. Some time during the night I was slightly disturbed by the entrance of a young lady—everything being dark I could not see neither did I care. She soon began to call Jenny! Jenny, but I did not answer. She felt of my bed, and soon began to undress. I remained quiet and being fatigued and sleepy I soon reembraced Morfeus in place of the young lady. On awaking next morning I found that my Bunky had left me but all her clothing and "other little unmentionables" I found by the side of the bed. None of the young ladies appeared at breakfast.

February 6

Hitched up and drove to Cedar [City] 18 miles where everything was arranged for them to take the stage, but when it came along the Major did not like to trust his family in such a rickety concern.

February 7

Hitched up and drove to Red Creek [Paragonah] stopped at Bishop [Silas S.] Smith's—a small room was assigned to us, a partition dividing us from the "sitting room" or rather the nursery, as I shold judge from the amount of juvenile music that was made there. Going to supper we had to pass through this room, but what a sight met my eyes, though interesting to me as a bachelor, but "what a sight for a father." I counted 21 youngsters of age varying from 6 months to 13 years having passed in review. I next entered the kitchen and dining room being a combination of one. Here I was introduced to three Mrs. Smiths, none of them being over 32 nor yet good looking. The head of this vast multitude and strict follower of a passage in the Bible "Multiply and all increase man" being absent on business [at] Salt Lake.

February 8

Hitched up and drove off for Beaver leaving the Smiths to enjoy themselves as best they could. Arrived at Beaver about 4 PM, having come 35 miles. Stopped at the Beaver House, kept by a Mr. Thompson, an apostate from the Mormon Church.[43]

February 9

About noon I hitched up and started for home but did not go far—only about eight miles—Major took stage.

February 10

Hitched up and drove to Red Creek to my old friends the Smiths whom I found still in a flourishing condition, but being unable to accommodate me with quarters, but being willing to allow my horses stabling, which was really all I cared for.

February 11

Drove to Cedar to Bishop Lunn [Henry Lunt]. Took in 500 lbs. flour and 300 [lbs.?] beans.

February 12	Started about 9 and drove to Dry Creek—camped out.
February 13	Drove to Toquerville. Stopped at Mr. Neberkers.
February 14	Hitched up and drove to the sheep troughs by way of Virgin City [Virgin, Utah]. Camped with Mr. [George?] Adair from Pa Rio Farms.
February 15	Started early, arrived at Short Creek about one PM. Camped the rest of the day. Solitair and alone.
February 16	Drove to Pipe Springs and camped at the Bishops.
February 17	Drove to Kanab from Capt. Bishops all alone, the rest of the party having gone to the "Buckskins." [Kaibab Plateau]
February 18	Resting.
February 19	About noon Prof. and his party returned. Packed up and drove to Kanab Gap and camped. Alfred Young joined party as herder.[44]
February 20	This morning I was installed as assistant photographer. Clem tried to take some pictures but failed, bath being out of order—fixed it—while manipulating he upset it—so much for the first day.[45]
February 21	Clem went to town for Beaman's old bath—fixed it up by Alteration.
February 22	Put up the dark tent—Clem put the bath in it open, started to a spring for some water about a quarter of a mile from camp. On our return, lo, our tent had blown over and the bath spilled again. Tried another which he upset a third time. While coming out he mixed up another. George Adair joined us as packer.[46]
February 23	Tryed our bath which worked all right. Packed up and started for Stewart Ranch. Camped in Oak Cañon by a small spring.
February 24	Packed up and started for the ranch. Began to snow which lasted all day. Got in about noon. A beautiful spring [Big Spring] starts from the side of the mountain about 250 [ft.] up and comes down on a slope of about 80 degrees. Water enough to run a turbin[e] wheel. Some enterprising man besides Bishop Stewart ought to

own it. Lumber could be sawed here to fence in all Utah. Magnificent. Pines measuring 125 feet in hight. Yellow pine.

February 25 — Still snowing. In camp. Clem and myself started down. Got three views.

February 26 — Tryed to take views but snow drove us in after taking two.

February 27 — Clem and myself started up the cañon. Took three fine views. Clem took two while I took one.

February 28 — Hunted horses which had wandered off during the storm. Failed to find them.

February 29 — Hunted horses again. Found them. Jones and Dodd started to find a way to the river across the mountain.[47] Andy and Fred started to the southwest point to build a monument.

March 1 — Dodds and Jones returned. Snow prevented them from going on. Some places 10 and 15 feet deep.

March 2 — Polished glass.[48]

March 3 — Clem and myself started with eight days' rations down the Kanab Wash. Took two pictures. Camped at the mouth of Oak Cañon.[49]

March 4 — Packed up and started about 10 AM. Arrived at the Wash about 4 PM. Camped at some water pockets.

March 5 — Started early. Made the head of Running Water about 5 PM and camped. Cañon walls begin at zero from the pockets and run up to 3000 ft. Almost perpendicular. Limestone formation.

March 6 — Rain this morning. Started to go to the River, being only about six miles from it, Bonny and Ryley being down there, washing gold, having left the party while I was off with the Major. But after having gone only about two miles it cleared up, so we concluded to go back, but on returning it began to rain again.

March 7 — Started back. Photographed all the best scenery. Camped at the head of the water.

March 8	Packed up and off early. Took views wherever we could get them. Camped near a pocket in the Rock.
March 9	Filtered our bath and started taking views as we went. Got to the Water Pockets about 7 PM, but oh, horror, our Pockets were empty—the rest of our party had camped there and emptied them—no water for us that night.
March 10	Up early hunting water. My search was rewarded by finding a large pocket full—what a Godsend it was. A man can stand hunger but not thirst—he gets fairly crazy. Had breakfast and started for Pipe Spring. Found some fine scenery—had to stop—took four views. Camped at the head of Running Water. Met Mr. Winsor and six miners. Winsor hunting horses while the miners were on their way to the River after gold.
March 11	Started early—got into Pipe Springs about two. Found mail for me from San Francisco and all the "Agles", Also William Johnson [William Derby Johnson, Jr.,] of Kanab who joined the party as assistant topographer.[50]
March 12	Commenced to fix a new Dark Tent.
March 13	Working at Dark Tent. The Deacon [Jones] and Johnson started for Wolf Spring to take bearings.
March 14	Finished my Tent. Deacon and Johnson returned.[51]
March 15	Commenced a Packsaddle.
March 16	Finished my saddle.
March 17	Made Sinches [cinches].
March 18	Tried our new Tent which proved a success.
March 19	Packed up for a trip for the Uinkarets Mountain. About noon a Mr. Fenimore [James Fennemore] reported to Prof. as Photographer, having been engaged by the Major at Salt Lake.[52]

March 20	Clem felt blue because he had to turn the instruments over to Fen. Told me that he would leave party as soon as his year expired. Fen. took picture of "Winsor Castle."[53]
March 21	Up and off early. When about half mile from Pipe Springs the pack on our mule became loosened. I called to Clem to assist me. While in the act of fixing it his horse scampered off, following the others. I told him to take my mule and follow. He soon returned with the horse, but his gun which was fastened to knob of saddle had been lost off. Clem remained behind to look for it.[54] Prof. and Fenimore started also back, but their search was in vain. They soon returned leaving Clem looking for it. Went about 14 miles and camped near a pocket. Saw three wild horses. Up to late—Clem had not come.
March 22	Packed up and off early—reached the foot of the mountain [Mt. Trumbull] about 5 PM. Hunted for water but could fine none. I pitied the poor horses. No coffee.
March 23	Struck an Indian trail. Packed up an[d] off before breakfast—followed it for about five miles which led to a large pocket full of water. The joy I felt I cannot express. Horses drank heartily. We soon prepared breakfast. Camped for the rest of the day. Fen and myself made views of the Pocket and Gulch.
March 24	Sunday being in camp, but Fen. took picture of the mountain. Called it "Mount Trumbull."[55] The Deacon and Fred went out to look for a Ranch [Whitemore Ranch] which was supposed to be somewhere about this mount—they fortunately found two men in search for stock from the Ranch. Prof. and Dodds went on the East side to look for water about 20 miles down. Fen and Johnson went to look for fossils. Mrs. Thompson gathered plants. Prof. and Cap. returned about 7 PM, having found water and also the Ranch.
March 25	Up and off early for the top of the mount[ain]. Prof. and his wife, Jones, Dodds, Johnson, Fred., Fen. and myself built a monument. After that all but Jones, Fred, Fen. and myself returned while Fred and the Deacon took bearings, while Fen. and myself tryed to photograph but it was too hazie. Stayed all night.
March 26	Begun to take views until noon, then went to the south end and descended, camp having been removed in that direction. After traveling about an hour and a half we discovered camp in an oak grove with a spring in the center.

March 27	Jones and Dodds started to find a way into the River.[56] Fen. and myself took three views of a lava bed—a recent outpouring. Returned at noon.
March 28	Took three views about camp. In the afternoon started toward the Colorado, had an awful time getting down to the foot of the mountains, it being all cut up with gulches and the lava being cut up into large ragged blocks. Took three views. Returned late at night. Could see the Cañon of the Colorado. Saw smoke near the foot of a black Hill near the cañon.
March 29	Packed up our mules and started back to where we passed down to the valley. Took two pictures. Found some Shenamo[57] hyrogliphics—tryed to copy them with the camera—failed on account of a storm coming up. Returned to camp about noon. Found Jones and Capt. Dodds home. Found that the smoke which I had seen yesterday had been their camp fire.
March 30	Jones and Fen. left for Kanab to bring rations and chemicals for photographing. Prof., Capt. Dodds and Johnson started to hunt a place to get down to the Colorado.
March 31	Snowing and raining alternately. Prof. returned this evening.
April 1	Andy and myself started toward the Colorado—took a picture of some Moqui hieroglyphics which were situated in a gulch near the foot of Mount Trumbull. Rode down to the River—got there about 5 PM. What a sight met my eyes! Looking down on the river from the top it appears to be nothing but a narrow gutter from the top to bottom—I should judge about 4000 feet and about 400 feet wide. It looked gloomy and forebidden. I counted eight rapids whose roar came up like a distant roll of thunder. Got to camp—found Andy waiting with supper. Camped by the side of a pocket.
April 2	Rain this morning. Could not make pictures.
April 3	Still raining. Our ration gave out. Took two views during the storm. Returned and got to camp about 3 PM. Found Capt. and Johnson had returned about sundown, having been successful in finding a road down to the River.
April 4	Snowing this morning. Fixed up to go down again but Prof. concluded not to being short of rations.

April 5 Packed up and started for Pine Valley. Snow in some places two feet. Lost the trail and started in a straight line for Berry's Spring. Made dry camp.

April 6 Started early but at ten it began to storm. Came to the edge of a line of cliffs—could not see the valley below. Unpacked and camped till one PM. Packed up and off—snowing at intervals. Made another dry camp.

April 7 Up and off early. Followed the line of cliffs all day. Finally came to a gulch which cut us off—perpendicular walls on both sides—could not find a place to get down in the valley. Camped near the gulch plenty of water. All we had was bread and coffee for supper, being our last.

April 8 Breakfast consisted of a pot of beans boiled in water no meat. Put out and struck for Gould's Ranch. Purchased some corn meal and molasses, then put out for Berry, where to our great joy we found our old friends Jones and Fenemore, also Clem and George Adair with lots of grub. My shot gun was also among the rations.

April 9 Found two letters and lots of newspapers. Commenced to fix a new dark tent.

April 10 Prof. and Adair started for Toquerville.

April 11 Fixed a dark tent and went hunting ducks. Shot three.

April 12 Packed up for a trip to the Uing karrets [Uinkarets] again, Capt. Dodds, Frederick, Fennemore and myself, Dodds to find a way to the river at several points, Fred to finish triangulating which could not be done during the storm and rations would not hold out till it might clear up, while Fennemore and myself went for pictures, which I failed to get on account of the weather. Camped the first night at Fort Pierce [Fort Pearce], a place used as a defense against the Indians but now abandoned.

April 13 Struck out early, took the wrong trail and went four miles out of our road, but finally found the trail. Kept along the cliffs. Camped near a pocket which we fortunately found.

April 14 Up and off early, still following the [Hurricane] cliffs. Saw four antelopes but too far to shoot. Dry camp tonight for the horses but plenty feed.

April 15	Still following the cliffs which go clear to the mountain. These hills, or rather the plateau on top, is called Hurricane Hill. Climbed the mountain and went to the east side to our oak grove camp, where man and beast drank heartily.
April 16	Fennemore and myself packed up and started for the river. Camped near the pocket where Hatten and myself had camped before. Day being dark and cloudy could not make pictures.
April 17	Started for the river. Tryed our luck but failed. Bath out of order. Started back and mixed up a new one.
April 18	Started down again, and made some fine views of the river.
April 19	For the river again. Obtained seven views. Fred and Capt. came down, having finished their work. Capt.'s Indian did not come to pilot him.
April 20	Capt. and Fred climbed down to the river. Fen. and myself started for more pictures. Made ten and returned to camp near pocket.
April 21	Packed up and started for our old camp in oak grove. Arrived about 3 PM.
April 22	Up and off early. Fred started for the ranch kept by Whit[e]more of St. George, in order to map the trail from the ranch to St. George. The rest started down by way of our old trail. After riding about four hours we met a man driving a mule loose while he himself carried the saddle, mule having broke his hobbles during the night, and was unable to catch him. We offered him help but the fool declined it. Asked him about the water, but his descriptive powers being on a equal with those of the "intelligent" contraband of the south, we could not find it where he said it was, or at least where we understood him to mean, but found a small seep in a gulch in the cliffs, the coyotes having dug it out, but it failed to supply us with the requisite amount, but concluded to stop and wait for Fred, but on retiring to bed had not come. I am very uneasy concerning his safety.
April 23	Up and off. Fen. and myself with the pack train. Exchanged animals with Capt. Dodds, my mule being a better traveler than his mare, him to go back to the ranch to look for him, we being off the trail about two miles, but on reaching the trail I found that he had passed—saw his mule tracks. I felt happy but sorry for Capt,

but still being in hope that he would follow the trail on which we had gone to the mountain. We started for a pocket 21 miles from our last camp at Coyote Springs, but on reaching this pocket all the water had left. I pittied poor Fred—no water since yesterday morning and no grub. He left a note that he had gone to the next pocket. Of course our animals, nearly perishing for water, had to go 12 miles more, and that might be dry too, for all we knew, for this pocket was full when we went up, but now not a drop in it. But on reaching it it still had plenty of water and the way the poor beast drank it I thought it would burst them. Found Fred had gone for corn 26 miles from here. Capt. returned at 11:45 PM, having found Fred's tracks after going ten miles, having traveled 55 miles in ten hours, and a hot day. Horses have no business in this country. Mules and jackasses are the only animals that can stand up in this dry country.

April 24

Started for camp by way of Fort Pierce. Stopped at the Fort to water and bait our horses and had a lunch of bread, coffee and jerk. Arrived here at Berry's Springs at 4 PM. Found mail from girls and [illegible] also a letter from John Doyle [Lee?].

April 25

Duck hunting this morning but got only one. All I saw. Wrote letters.

April 26

Shot two ducks, finished my letters. Jones went to Toquerville with the mail. Cleaned glass and fixed up the traps for a photographic tour to Mt. Kolob.

April 27

Broke up camp. Prof. and wife, Fred, Adair for Mary's Mt.,[58] Jones and Johnson for the Pine Valley Mts., Capt., Andy and Alf. for Kanab, Fennemore and myself for Kolob [Plateau]. Left about noon by way of Hurricane Hill and camped for the night at bank of the Virgin about half mile below town of that name [Virgin City]. In the evening we visited the bishop of the place, a Mr. [John] Parker. What a relief to the eye to look upon an orchard all in bloom after being out in the sage brush for a number of months. He seems to be lost, he is fairly enchanted. Scenes of childhood, long forgotten flash up before the eye of memory in vivid succession—how when a boy he looked forward to the time when those trees, now in bloom, should be heavy laden with fruit—how he would watch day after day until fit for digestion. We spent a few hours with the Bishop and his wife in pleasant conversation. Nice folks they. Returned to camp, and consigned ourselves to Morpheus.

April 28

Packed our mules and started for the Mt. Called in at the Bishop's whom we had promised to make a picture of his house, but on account of the high wind and clouded state of the weather, we were unable to fulfil our promise. Procured a canteen of wine, the sample last night being very good. Off we started for the top, following a trail which led up a creek very rapid and shallow, called by the natives North Creek. All along the bottom for seven miles where a small spot free from rock appeared, it was under cultivation, mostly vineyard. The grape seems to flourish here. This southern Utah is called Dixie. Near the head is a saw mill. Lumber is made here for all those who haul the logs, which they get from the top of Kolob, a distance of ten miles. Having climbed some 3000 feet we encountered a snow storm and my old soldier came into requisition. What a change from four hours ago, people walking about in their shirty sleeves. The scenery is grand of its kind and affords fine subjects for the camera. The formation is homogenus sandstone. The top had been overflowed with lava. Some parts of a snow white, while others of it a dark red and intermediate color. Pinnacles innumerable, forming an immense harrow upside down. Here and there are scattered peaks, overlooking the others by a thousand feet, or like giants among dwarfs. All bare rock, no vegetation on the towers. Camped near a little seep spring.

April 29

Up and off for views; after going about three miles we came to a house half finished and a beautiful spring. Stock is kept here during the summer. We unpacked and tryed to make pictures but failed on account of the high wind and storm of snow. Camped in the house. Toward evening Fen and myself took a walk toward the head of a cañon, climbed down about a thousand feet through brush and rocks. Found a big spring running out from under a bed of lava. Preferring to climb up the vertical wall of lava to scrambling back through the brush we began our task, but before getting half way up we felt sorry for our bargain. Reached the house quite late. Was surprised to see a big fire in the house, but on nearing found a visitor in the person of a Joseph Gibbs, of Dunken's Retreat,[59] in search of stock.

April 30

Out early on the hill. Made ten pictures, then returned to dinner, bringing with us our traps. Had an awful time catching our animals. After dinner packed up and started to the Twin Peaks of pure white sandstone. Made two views and returned. Our visitor had left us, but what a fright—the bars which we had put up in front of the door had been removed, but on nearing found that they had been thrown down by some ferocious brute. All our cooking utensils was buried in the mud and sand. A pound of fresh butter which we had left on a plate could not be found at all—probably eaten for the salt. One spoon and fork could not be found. Our negatives and chemicals were arranged along the wall which miraculously had escaped his wrath. Probably the smell prevented him from molesting them.

May 1	Packed up and started for an Alcove Cliff. Took three views, then returned to our first water. Made three views and camped, hungry and weary from our day's work.
May 2	Packed up and started for Virgin City.
May 5	Washing glass. Jones and Johnson joined us.[60]
May 6	Polishing glass. No party.
May 7	Cleaning glass.
May 8	Washed my clothes.
May 9	Party returned at 9:30 PM.[61]
May 10	Packed up and started for Kanab. Camped just below town.
May 11	Had to cook, that functionary [Hattan] having gone to the mouth of Paria. Fenimore fixed up for printing pictures.
May 12	Stopped cooking and went to work with Fenimore.
May 13	Jones took a team and started for Pipe [Springs] to bring the rest of our stores.
May 14	Still printing. Jones, Andy and Capt. returned this evening.[62] Weather cold and rainy.
May 15	Printing finished this day.
May 16	Mounted our pictures.
May 17	Still to work mounting.
May 18	Fixed up to start for Dirty Devil.[63]
May 19	In camp fixing up.
May 20	Still fixing.

May 21	Made Clem a dark tent, to[ok] a feriotype of Fred.
May 22	Still in camp.
May 23	Everything ready for tomorrow's start.
May 24	Prof. decided to wait till tomorrow.
May 25	Off this morning Eight Mile Spring. Found no water. Off for Johnson Cañon. Camped about 2 ½ miles above settlement.
May 26	Sunday. Taking it cool
May 27	Hitched up and started for Kanab with the wagon, returned same day, bringing with me George and Fen. who had remained behind. Fen. forgot his filtering tank, returned for it but failed to catch up. No sign of him up to ten PM.
May 28	Fenemore came to camp about noon, having got lost during night. His mule broke away and so had to foot it. A party of prospectors came from Salt Lake. Camped with us. Willie Johnson paid us a visit.
May 29	Fenemore went back after his saddle—returned at noon. Prof. came up from Kanab about 9 AM. Prof., Fred and Johnson went up on point "B" for observations. Returned at night bringing Fen's mule which they found with a band of Johnson's horses. Alfred Young started for Kanab. Struck for more wages which was refused and so Prof. discharged him.
May 30	Got our pack ready, started about 10 AM. up the cañon, reached Lee's Ranch about 4 PM.[64] I understand from George that J.D.L. has sold out for 1800 [dollars] to a man by the name of [John Wesley] Clark. One of Lee's women [Lavina Young Lee] still resides here, while he has another with him at the Paria.[65] This is a splendid place for stock. Fred and myself climbed up a little butte. Fred sketched the surrounding cliffs and valley.
May 31	Started early, struck across the country in a northerly direction. Towards evening struck a nice little valley [Adair Valley] with a clear cold spring [Adair Spring]. Concluded to camp. After supper went out to hunt. At the lower end of the valley found a beautiful lake. Saw lots of ducks. Killed three. The rest hiding in the reeds which

surround the lake. Adair put his preemption stake at the spring. Called it Swallow's Park, and Lake Adair. Indian guided us across.[66]

June 1

Off at an early hour. Struck an old trail which is supposed to be the trail of a hundred men who started from Kanab in '67 [1866] on an exploration of the country in general, all Mormons, a man by the name of Jim Andrews [James Andrus] in charge. After winding and twisting up and down hill for 16 miles we went down a large gulch containing running water. Our attention was attracted by two tombstones but the grave had been dug up either by Indians or wolves as the bones were scattered about. This grave it is supposed contained the body of Everett, [Elijah Averett] one of the 100. This Everett and seven others were sent back with fatigued horses to Kanab. While going up the hill Everett was first, a couple of Indians having tracked the party, seeing these return, fired on them and killing Everett. The rest being none but boys fled back, the Indians carrying off their horses. Had Everett followed the instructions of Andrews his life would have been saved. Among the band of horses which they took back was a mare who could smell Indians a quarter mile off and could not be got to go up to one. The instructions from Andrews were to drive this mare ahead to make sure, but they led her in place of driving her. One of the boys returned to camp, told his story, when the party gave chase recapturing the horses and guns. The Indians fled to the mountains, which afterward were captured and roasted. Camped at Pa Rio [Paria] River.

June 2

Went up the River to its head.[67] When about 12 miles from camp I recollected that I had forgotten my ammunition. I returned, found it all right—was overtaken by a regular mountain thunder shower which saturated every stitch on me. I followed the track of the party which I found had gone over the divide into Potatoe Valley. Climbed up on a narrow strip resembling a hog back, from bottom to top 2500 feet—in some places it was only 2 ½ feet wide on top. The sun was just going down. When I reached the top I stopped to breathe. I looked down the awful chasm—it looked wild and forbidden, the valley below being in deep shadow. Table Mountain [Table Cliffs] north of it with its pink colors reflected back the golden light of the setting sun long after she had gone from my view. A wilder view I never beheld. It surpassed the Wild Scene on the Colorado at Lava Falls. What a picture for Burstadt [Bierstadt][68] After leaving the edge of the cliff I found myself in a beautiful little valley hemmed in by mountains on each side. Green grass dotted with flowers—in the distance I saw the blazing camp fire of my companions who probably feared I might not reach it, as night set in very dark, but my faithful little mule never missed a step from the trail, having come 42 miles. Supper was eaten with a relish which only a mountaineer enjoys. After supper a smoke of "Navy", after which I joined the chorus of snorers.

June 3

Our Indian felt sick and wanted to go back. Prof. gave him an old horse blanket and some flour then "Pike way'd" for home or some other place. We followed down the valley. This is the head of the Dirty Devil River[69] it starts by springs. At night found a large creek [Escalante Creek] coming in on the left, where we camped.

June 4

Rained all night and all day long. Prof. feeling sick concluded to rest.

June 5

Packed up and started down the valley about 8 miles. Here the river ran into a narrow cañon which made us halt. Unpacked—had dinner. A tilted ledge about a thousand feet high runs square across the valley running northwest and southeast a quarter of a mile in the cañon—a rapid little creek comes in on the left, coming down from Lake Mountain [Acquarius Plateau] who loomes up in the distance with his snowy garb and tapering pines. After dinner everybody climbed out or looked for a short cut to the mouth of the Dirty Devil River. Fen. and myself climbed up the ledge with instruments to make pictures—succeeded in getting two fine ones showing the junction of the two streams and also the cañon. Returned to camp. The General soon after spread the cloth for supper. Capt. and Prof. made their appearance on the hill and soon joined us at supper. The[y] reported having followed a trail twelve miles leading to a gulch with plenty of water pockets.

June 6

Packed up while Prof. and Capt. went off on the trail trying to follow it. I brought the train to the gulch [Harris Wash] and camped. Fen. and myself started down the gulch for some pictures. After going about a mile we came to a jump down of nearly a thousand feet deep, while the walls above us nearly measured the same. It gradually tapered down making it look like a huge bin, top of cañon being nearly 2000 feet wide while at the bottom on[ly] six feet, no water being in it, only in pocket. Photographed it and two other views, and then returned to camp. Found Prof. and Capt. had returned. Prof. reported that he was satisfied that this creek which we had been following was not the head of Dirty Devil River—what could it be—but finally came to think that a small stream came in on the right just above the San Juan River—being now about only 50 miles from Colorado River, for Mt. Seneca Howland [Navajo Mountain] being in plain sight, and the whole country seemed to sink towards it, while the Dirty Devil Mountains [Henry Mountains] loomed up 75 miles north of us and the river flowed on the northern side, so it was impossible for this stream to take a turn and flow up grade.

June 7

Prof., Capt., Clem and Adair went to a high point to obtain a panoramic photograph of the country while I took packtrain back to our old camp which we had left yesterday morning. Fen. and myself, with animal started down the cañon a little ways for pictures but could not get down a great ways, for where the two creeks join their water(s) form a deep stream, especially this season of the year. The creek coming down from the west is called Lake Creek and the other Birch [Creek]. Made three pictures then returned. Found the rest of the party had also arrived. Soon spread the cloth and "went for it," I mean supper. Prof. had concluded to send three men [George Adair, S.V. Jones, and W. C. Powell] back to Kanab for Rations and return to this place as soon as possible. I was designated to bring the boat down, also Fred, Johnson and Fennemore. I hastily scribbled a few lines to my friend Dick Geits.

June 8

Adair, Clem and the Deacon started off for Kanab with seven animals, while we packed nine and rode seven. Started up Lake Creek toward Lake Mountain but left the creek near mountain to our right where the cañon followed a Gulch at the foothills for about six miles. Stopped for ¾ of a hour. Prof. and Capt. climbed out but soon returned and reported that we had to climb up—up we went 2000 feet—a grand sight is had from the top. About sundown we passed around a mountain—the scene was grand—it put me in mind of crossing the Isthmus of Panama. Struck a beautiful little Creek [Boulder Creek?] which came rushing taring down from the top of the crest. Its sparkling cold water was enjoyed by both men and beast. Feed the best I have had since leaving the States. Camped for the night.

June 9

Packed up and off. This mountain travel is fearful—no sooner have you reached the top of some Giant than you will have to climb down again on the other side some thousands of feet. Reaching the bottom or rather the divide you find a cañon which cuts you off—now you have to go up to its head. Perhaps you are a lucky Dog and find where the water has undermined the bank and let it slope down at an angle of 60°. Now put on the corsetts, but no matter how tight you make them something will give way and away goes your pack. You're lucky if you find everything together again—some is sure to reach the water and then that is the last of it, for the water rushes down with a tremendous force and sweeps it beyond your reach. Followed up quite a stream for about five miles—it ran down at an angle of 30°. It came down leaping, taring, rushing, now running down a smooth slope, now taring over big boulders, now leaping down 50 or 100 feet at the bottom of which it finds itself churned into white foam—here it collects itself and again is off, never stops until it finds itself in the Pacific Ocean, via Colorado River. We crossed seven beautiful streams—all along these streams the finest pasture I saw in Utah. Winding around a huge old Mt. all of a sudden we saw a beautiful lake [Aspen Lake] about half mile wide and the same in length. Fenemore and myself made a picture of it. After supper went down and shot several ducks.

Poor Fen. is sick this morning with cramps in the stomach, but he rode off with us. Stopped at noon by a roaring stream. Before reaching noon camp Capt. climbed up a tree and reported two lakes east from our trail, also reported a beautiful valley, for which we started and nooned. Prof. concluded to remain here, Fenemore feeling too sick to travel—Fred and Johnson went with me to the lakes, which I photographed. Called the "Bee" Lake. Fred called the other Hidden Lake. The finest place for a ranch I have seen in Utah is right here. This valley including some low hills, that is, the foot hills of the Lake Mountains, is 20 miles square, with streams running down every five miles. Feasted on Pine hens.

Long time finding our horses. Did not get started till about 9:30. After filling themselves with the long green grass they had taken to travelling; about two miles from camp we struck a stunning creek—it ran the most water of any we have seen—it ran very swift, jumping and foaming as it went down the steep slope, full of little falls called it Cascade Creek from the top of the mt. It leaps over a perpendicular cliff over a thousand feet high, but its fall is broken by a large rock sticking out from the side about three hundred feet from the top—the rest of the way it goes a little slanting at an angle of about 80°. Did not go close to it for our time is precious, and rations short. Crossed in all nine creeks of different sizes, all running into Burch Creek and then into the Colorado. In the afternoon we left the southern slope of the Lake Mountain Range and crossed to the east side. The country below us is all cut up with gulches and cañons for miles—nothing but sand rock is visible. The Dirty Devil Range [Henry Mountains] from this point looks like a dry country and almost impossible to get to them. The first peak is about 30 miles from this point eastward.

Prof. and Capt. went in a northeasterly direction to find a trail or hunt up the best place for to get to the mountains near which our river flows. Fred and Johnson went southeast on a similar errand. Fen. and myself went down a cañon due east for pictures. Secured three—a storm coming up we returned to camp. Shot a dusky grouse; on our return found Fred and Johnson home. They reported having found a heavy beaten but not very old trail leading in the direction of the mts. About 5 PM Prof. and Capt. returned. Reported having seen fresh Indian sign. Rained very hard.

June 13

Followed the trail F. and J. had found which led us through small valleys and cañon and to wind up, lead us down from the plateau on smooth sand rock over a thousand feet high—horses often slipped for a number of feet. Sometimes found ourselves on projected shelves, not over two and a half feet in width, but got down all OK. The last turn brought us in sight of a beautiful valley running northeast [Pleasant Creek] while from the west came a rushing stream which flowed in a small cañon of sand. All along the edges the cottonwood trees flourished. Camped in the cañon. Plenty feed for our animals—wild oats in abundance. Followed fresh Indian tracks all day.

June 14

Up and off early. While making a descent from a Bench we were attracted by the bark of a dog. Looking in the direction from hence we heard it we saw two squaws flighing through the grease wood, yelling as though they had seen so many devils. We saw their camp on a small hill for which we started. On nearing an old man about seventy met us at the foot of the hill, trembling like an aspen leaf. On reaching this camp found it deserted. Guns, bows and arrows had been hastily left in their wickiups, but we soon dismounted and seated ourselves around his camp fire, where we allayed his fears by telling him to smoke, at the same time handing him tobacco. At this token of friendship he steadied his nerves and began to talk. Found him to be one of the Red Lake Utes down here to gather seeds. A quarter of a hour after in came the two frightened squaws, who began to chatter like monkees. [A] little while after we saw two young men across a gulch on a small hill. The Squaws began to call to them not to be afraid for we were Tuitchea Ticabu (very friendly). They soon came slowly toward us looking like shamed men—no doubt they felt so. After smoking and talking, They begged us to go into camp and trade with them. Prof. being anxious to know about the country at large, and all about the trail, he consented to camp—traded several buck skins. I obtained a splendid skin for a small paper of paint.

June 15

Struck off on a trail up the valley—followed it into a cañon. About a mile from its mouth we lost sight of it with the exception of a track here and there—probably cattle feeding. This is the great hiding [place] of the Indians and many heads have found their way in here, all stolen from the Mormons, who never suspected for a moment that their friends the Utes would do the like, but thinking the Navajoes the guilty Party. The Utes always watched the opportunity, when a band of the former were in the settlements. The Mormons could track their stock for quite a ways, but as soon as they got into a sandstone country they gave up the chase, being impossible to track them over the bare sandstone, and no one thought it for a moment possible to get down to the valley below. They never dreamed that here the Indian feasted on broiled steak. Wild oats grow here the same as cultivated does anywhere else but not so heavy. At present is the time when the Indian gathers his yearly supply of seeds and nuts. Those which we left had quite a crop gathered. I have no doubt that our party is the first white party here. The Indians felt surprised how we got in—asked numerous questions—how we found our way in. Wandered about all day trying to find a trail which would lead us out of the cañon besides the one we came in by without success. The cañon walls are perpendicular from 700 to 1000 ft. in height.

June 16

On the search again. At noon Prof. concluded that four of us should climb out and head all the gulches. Prof. and myself went one way while Capt. and Fred went the other. Found a nice level Platteaux on top, but no sign or a trail, but found some very old horse dung, probably three or four years old. After heading all the side gulches we walked towards the mountains. Here we found ourselves on the divide. We could see the break of the Colorado plainly and only about twenty miles distance. Half the south side of the mountain is drained by a large cañon flowing east to c. [Colorado] while the other half flows west and then north into the Dirty Devil River which flows on the north side east into the Colorado. Having studied out our course we were bound to get out of the cañon. On our return we climbed down one of the side cañons and here we thought we could get up with a little work, but rather hard for the horses, but we were bound to get up. On our return about 4 PM found Capt. and Fred in camp, who reported having seen no sign of a trail, but had found about three miles from where we climbed out, two water pockets, and as our camp was about that distance from the water in the cañon, we concluded to try our cañon wall. After two hours' work we reached the top. One horse fell backward while going up a steep ledge and fell about ten feet, but picked himself up, shaked himself, and tryed it again—that time succeeded in reaching the top. Camped near the pockets.

June 17

Got off about 9 AM. Struck out on our trail from yesterday, reached the foot of the mountain [Mount Pennel] about 3 PM. [The] foot hills are heavily studded with Pinone Pine, only found on small streams of water coming down. Climbed up to nearly the top and camped at the head of the spring. Water is very cold. The mountain is ignius rock, Trachyte. No heavy timber grows on it. Fir, aspen, oak (shrub) and pine—both kinds. A very disagreeable day, cold and windy.

June 18

Remained camped. Prof., Capt. and myself climbed the second mountain [Mount Pennel] from the north end. Fred and Johnson climbed the north mt [Mount Ellen]. After two hours' climbing found ourselves on top. Our Barometer read 13000 feet above the level with the sea, and 3000 above our camp, this being about half way up from the foothills. A grand sight is obtained from this point of the surrounding country. Far to the north is the famous Wasatch Range, while the Sierra La Sal looms up NE from here, and the Sierra Abajo lies due east. Looking south, Mt. Seneca Howland [Navajo Mountain] stands like a sentinel guarding the junction of the San Juan and Colorado, and west is the home of Truthfull James, "Table Mt." The intermediate country between here and all these mountains is cut up by cañons and gulches. It seems almost impossible for men to travel over it. While on top a heavy cloud came over, and a snow storm was the consequence. After this had passed we could see another cloud below us—it looked grand. After satisfying ourselves about the locality of the mouth of the Dirty Devil we made our descent. Got to camp in about half an hour. Fred and Johnson returned about sundown.

June 19

Started early. Crossed over the divide to the north side between the first and second mountain, and then east. Passed through a forest of cedars. About two o'clock went into camp on the bank of a beautiful little creek [Trachyte Creek] coming down from the second mt. Prof. and Capt. went out to see which was the best way to get into the River, which from this point is plainly visible, but the section of country between here and the mouth of the Dirty Devil is fearfully cut up. Prof. and Capt. returned at supper, having found a trail.

June 20

Started off on the trail which led back again to the little creek and followed it down for quite a ways, then taking its course across a Platteaux, which was very sandy. Here we lost all traces of it. We started down the creek again, which toward evening wound too much to the south, while our course was east. Went into camp. Prof. and Capt. climbed out but soon returned, having studied out directions, and decided that they would start early tomorrow and find a way in, and if possible to find the gulch in which Prof. had climbed out from the River last fall.

June 21

Prof. and Capt. started early. Capt. returned about noon—reported having found the gulch, but a Devil's ladder to go down on. Being already, we skinned out, and after three hours travel we reached the head of the gulch [Crescent Wash] which from the top looked dark and gloomy below. After trun[d]ling several pack horses down the cliff, making their hides look like the map of Spain, we finally reached the bottom. Went into camp in a cottonwood grove nearby. Found a pocket of water.

June 22

Off early. Wound our way down the gulch—reached the River at noon—found it booming and chuck full. After dinner started for the mouth of Dirty River on foot, a distance of three miles. Here we found our boat, the *Cañonita*, all right but the water had washed her sides and floated one oar part way off but lodged between the rocks. We soon cleared her cabins and pulled her out. Soon had her caulked up and went to camp, easier than walking along the cliff. The distance from the water to where we housed her last winter, or fall rather, I estimated at 40 feet, but at present was ten feet below the boat, though it had, as above mentioned, washed the oars, which were buried under her sides in the sand. On reaching camp we hauled out and turned her bottom up, washed her off ready for general repairing and painting.

June 23

Caulking boat. Prof., Capt. and Andy started back at 5 PM. I wish them a safe journey back.

June 24

Finished caulking and put a coat of paint on her.

June 25

Put another coat of paint on her. Fixed up our rations.

June 26

Put our rations and other traps in. Started down on swift current—made four miles—stopped to take pictures of an old Moqui Ruin it stands on a cliff about 15 feet high, 18 feet wide and 20 long.[70] The front, facing north, has fallen in. It must have been hundreds of years ago since it was built. The walls, that is, the stones were honeycombed by age. A good deal of taste was displayed in building. Each room was squarely laid up. As a cement they had used a mortar of mud which was only perceptible at intervals and in thin layers. Camped for the night. Made four views.

June 27

Started about 9 AM. Nothing of interest presented itself to photograph. We ran 9 miles and camped Here we photographed more ruins[71] made on one nagitive.

June 28

Run down to Pine Alcove Creek—made two views—wind blowing a gale. Found that we had forgotten to bring trypods for dark tent, also focusing cloth was left behind. Rowed to the left bank. Fen. returned. He climbed over the cliff, being a cut off. Distance run today 1 ¾ miles. Fen. returned about supper time.

June 29

Made three views, then started down. Went about five miles. Stopped on the right, made two pictures. Had dinner—ran about a mile—stopped—made two pictures of Red Bud Cañon. Ran down about a mile and half more and camped at the mouth of creek coming in on the right.

June 30

Pulled out and ran about two miles and a half, landed on the left. Fred and myself climbed out—found two large pockets about 10 feet wide and 20 deep. Wind blew a perfect gale—could hardly keep on our feet. Laid up for the rest of the day. Could not photograph on account of wind.

July 1

Went up a side cañon—found a large water pocket at the head. Made 9 negatives. Returned to camp and dropped down about two miles. Camped on the left. Two cañons coming in—one flows a good sized creek, the other dry, with large pocket at the head. A large island in the shape of a heart is just above camp.

July 2

Started up the dry cañon. On going to the head of the Cañon found an old decayed Moqui ruin situated under the over-hanging cliff. It was about 70 feet long and 20 wide. I climbed up to it. Found four sticks of oak, but in touching them [they] fell to pieces like Rip Van Winkle's gun. Made two pictures of cañon and one of the Island, and two others. Returned at noon. After dinner dropped down about ten miles. Saw a cañon coming in on the right. We stopped, Fen and myself took a walk up. I noticed pottery and arrow chips lying around, so we began to look for ruins. We discovered a square hole in the wall of the cañon about 12 feet from the ground, but a ledge ran up to it. On examining this closer we found it the work of the Moquis. They had walled up a large cave, leaving a hole in the center two feet square. I entered, found an old corn cob. No doubt they had used this as a store house. Below we could just trace the foundation of three houses. Found the butt of a spear.

July 3

Went up and made a picture of the house and two of the cañon. Returned, had dinner and pulled out. About a mile down we stopped and made two pictures of a vertical wall, beautifully frescoed. Ran down about two miles more, where Fen. wanted to take a picture but could only get it in the morning. We camped on the left. Fred and myself climbed out. Could overlook the country for miles. Looking northwest, saw a beautiful valley coming in, and also a creek, its edge studded with cotton woods.

July 4

Glorious Fourth again finds me in the Cañons of the Colorado. Fred, as last Fourth, fired a National Salute. Fen and myself made two views, returned at noon—found a layout—a peach pie and coffee cake. It was excellent. Cake light as a feather. I doubt if anybody in the whole United States enjoyed their dinner better than we did, though our cake and pie was baked in a frying pan—but necessity is the mother of invention. Fortunately having two of these gravy makers, Fred made an oven out of the two, turning one on top of the other, in which he put his coals. After dinner, all hands for the top of the cañon. Fred sketched in the country, while Fen. and myself made negatives. Succeeded in getting four. Being very hot, we could hardly manipulate. Had to give it up. Returned to camp about 5 PM.

July 5 Ran down about three miles and made two pictures. Ran two miles farther to an alcove on the left, made one picture. Had to give it up on account of the high wind. Ran down a few miles more and went into camp. Cleaned glass for the rest of the day.

July 6 Made four pictures, then ran down to an alcove on the left. Made four more—ran down to the long cliff bend and camped.

July 7 Made four pictures, then ran down about four miles and made four more. Ran four miles more, passed the mouth of the big Boulder [Escalante River], but here it flows only from eight to ten inches of water, while on the mountain it just boomed. The sandstone drinks it all. Camped on the left.

July 8 Ran down half a mile. Made three pictures—packed up and off: Stopped at the junction of the Colorado and San Juan rivers. Made two pictures. About a mile below the junction, camped for dinner. Fred and Johnson tried to climb out but failed. Returned about 4 PM. Ran down to the Music Temple. Could make no pictures on account of light being bad. Monument Cañon [Glen Canyon].

July 9 Rained all day. In camp.

July 10 Still raining. Cleared up toward evening. Made three views.

July 11 Finished up and ran down to Mt. Howland [Navajo Mountain]. Stopped and made three pictures. Ran a rapid OK. Had swift water for 15 miles. Camped on the right.

July 12 Off early. Ran down to crossing [Crossing of the Fathers]. Dug up a cache of photographic material. Had dinner and pulled out for four hours. Made 25 miles. Camped early on the left. Passed several creeks. One is called "Sentinel" [Wahweap]—a huge column of detached rock stands at its mouth like a sentinel.

July 13

Ran down to the mouth of Pah Rio in an hour. Here we expected someone with rations for us, but no one here. Knowing that J. D. Lee had one of his wives [Emma] living here, we started up—found the old gentleman [Lee] in the field plowing. After stating our case, he told us to make his home our home until our men came down, which we accepted. Gave him some flour.

July 14

No one yet. Helped Mr. Lee on the road, who started for Jacob's Pools. Willie Johnson went with him.

July 15

Hoed onions and beets in the forenoon. Had just returned to the boats when we heard a shout, which afterward proved to be Clem's, who told us that Andy was down the river with the wagon and wanted us to come and help him over a bad place. All went down and helped him up. Read all afternoon.

July 16

Fixed up a bridge across the Pah Rio.

July 17

Carted over some of our provisions and went to work at the *Dean*.

July 18

Caulked and painted.

July 19

Painted.

July 20

Painted.

July 21

Worked on Lee's dam.

July 22

In camp reading.

July 23

Cleaned glass.

July 24

Mr. Lee invited us to spend the day with him, being the anniversary of their entrance into the Salt Lake Valley. Had a splendid dinner, played cards and sang songs. After supper returned to camp without a change of opinion of Mormonism.

July 25

Cleaning glass.

July 26

Made a ground glass for Clem.

July 27	Fennemore feeling very sick.
July 28	Fixed Clem's tripods and camera box.
July 29	Hard to work trying to kill time.
July 30	Doing about the same.
July 31	Spent the day with Lee.
August 1	Sleeping and reading.
August 2	Commenced to make a cultivator for Lee.
August 3	Finished it.
August 4	Concluded for someone to go to Kanab and find out true state of affairs, and also to report Mr. Fennemore unable to go down the river.[72] Held a counsel to see who should go. Concluded to draw lots. It fell on Andy, but afterward Fred took his place. Clem and myself started up for the horses. Found them seven miles up the cañon on the Pah Rio. Returned about 4 PM. I let Fred take my mule—after supper he started, taking with him our mail.
August 5	Trying to kill time.
August 6	Spent the day with Lee.
August 7	Reading and sleeping.
August 8	Same as yesterday. Raining.
August 9	Saw some geese up the river, pulled over, but when we got there the birds had flown. About 6 PM heard our signal. Hastened down the river. Found Jones, Fred, and Lyman Hamblin with a load of provisions stalled—helped them to make a portage.
August 10	Received a lot of newspapers. Boys reported the Uintah Utes on the war path.[73] Johnson gone to Salt Lake. Capt. [Francis M.] Bishop driving team for Major. Received a box of candy for the "Boys" from Steward. Three cheers were given.

| August 11 | Spent the day reading and packed chemicals. |

| August 12 | No party yet. Reading and sleeping. |

| August 13 | Day spent the same as usual. About 5 PM heard our signals. Answered and hastened down. Found the Major, Prof. and wife, Prof. Du Mott [DeMotte][74] and George Adair. Indian Ben for a guide. "Quawgunt" [Kwagunt][75] In the evening Jones, Fred and myself took Mrs. Thompson and Du Mott boat riding. |

| August 14 | Sacked up the rations, put them on the two boats, the *Dean* and *Cañonita*. After dinner ran the rapids just below the Pahrio. Mrs. Thompson rode over with us. Land just at the foot of it and went into camp. |

| August 15 | Helped Clem to make a picture then went up to Lee's. Helped to catch horses and mules for the returning party. Signed four vouchers for Major. Had dinner at Lee's Party got started about 2 PM. Mr. Fennemore felt sick. I pity him.[76] |

| August 16 | Put on extra planking along the keel of both boats. Beaman and [James] Carlton passed here on their way to the Moquis.[77] |

| August 17 | Major concluded to take only one photographic outfit. Cached Clem's at Lee's, to be brought to Kanab at the first opportunity. On our return to the boats we started down Marble Cañon. Walls are very low, but run up very rapidly—grey sandstone at first then the red. We soon reached our old friends the rapids. Ran over them until we had passed seven. Camped on the left for dinner and pictures. Walls are now from 7–800 feet high. After dinner we have a portage then ran four more rapids. Camped on the right at the head of another portage. |

| August 18 | Made the portage. Made a picture of it from below. Camped for dinner at the head of a rapid on a ledge of rock. No talus. Clem and myself made negatives. Burnt my foot badly. After dinner pulled out the rapid, from that into another which proved to be a portage. After that we had a let down. Camped on the right. Walls 1200 feet—grey and red sandstone with a bedding of limestone. |

| August 19 | Commenced with running two rapids, then came to a portage. Had dinner at the foot. Made a few negatives. After dinner ran two big rapids and let down three more. Camped on the left. Rained all night. |

August 20	Broke camp 8 AM. Ran a buster of a rapid all OK. Came to a portage, made it, then ran rapid after rapid. Camped for dinner at the head of a big rapid. Made pictures. Large gulch came in on the left. After dinner ran the [illegible] rapid, shipped a good deal of water. Camped early. Made some pictures of a side cañon. Have been running the marble up and is now up 100 feet. Walls of cañon 2000 feet. The *Cañonita* got full of water, and Prof. not having securely tied the bag in which our negatives were kept got wet and were spoiled—threw them away. Made one portage and ran 12 rapids.
August 21	Pulled out this morning at 7 ½. Ran down a mile and stopped to make a picture of a spring coming out of the side of the cañon wall—flowed quite a stream. Also made a picture of a cave. The river, in making a sharp bend, had undermined the wall over 500 feet. Pulled out for a mile or so, then camped for dinner and pictures. Pulled out after dinner. Ran two [of] the biggest rapids we have seen. Cañon very narrow, walls 3000 feet, nearly vertical, no foothold for portages. Camped on the left on a sand bank. Ran 10 rapids.
August 22	Made two negatives before starting. Ran five rapids. Stopped on the left for a few pictures. Stopped for dinner and pictures on the right bank Found a field of cactus apples, very delicious fruit. Have run the marble up. A sort of greenish looking shale is making its appearance. Walls 3300 feet. Vertical on top and talused below. Two gulches coming in from the Buckskin [Kaibab Plateau]. Could see it loom up some 6000 feet heavily studded with timber. Pulled out, ran a rapid, then had a line portage. Kept on running rapids. About 4 PM came to the mouth of the Little Colorado or Flax River, an awful muddy stream and so salt that it cannot be used as a thirst quencher. Camped for time and latitude. This is the end of Marble Cañon, 62 miles in length. Ran 63 rapids, made four portages and let down by line five times.
August 23	In camp. Clem and myself making negatives. Fred climbed out. Prof. and Jones taking observation found a fire place under the rocks, probably built by the Cohonies [Cohoninas].
August 24	Broke camp and pulled out into the Grand Cañon of the Colorado. We began with running a rapid which was followed by four others. When about 6 miles down came to a fissure from which had flowed lava, damming up the river, but the river has cut through it again, only leaving a fall. Went into camp. Major went out to

geologise. After dinner Clem and myself went up the river for pictures. Right wall of the cañon is 6500 feet high.

August 25

Still in camp. Rowed up the river for pictures. Returned at noon. Made a few more about camp in the afternoon. Major went across the river—found a silver vein in the fissure but how rich he could not tell.

August 26

In camp. Party climbed out again. Major studying a fault and hunting fossils.

August 27

Major and Fred climbed for a couple of hours. Prof. filling barometer tube. Broke camp at 10 AM. Ran over the fall without shipping much water. Ran 9 more and went into camp on the left. All hands climbed out except Andy and myself. Jones, Major and Fred rowed across the river and climbed.

August 28

Pulled out early. Ran five rapids then came to a buster. Let down by line, made it, pulled out, ran another. Landed on the left. Made two fine negatives. Had dinner and pulled out into a rapid, then came to a let down. Ran another small rapid. Came to a young Hell, unloaded the boats. Here it commenced to rain. Being wet from head to foot anyhow, but had to hunt shelter from the beating rain. Got in the river under the side of the boat. Rain over, we hauled the boats over. In doing so I sprained my back. I sunk down, cursed my luck, and crawled to shore. Went into camp—everything wet. Laid down on the wet sand on damp blankets.

August 29

No sleep last night. Could not raise up. My back seemed broken. Short of rations. Had to do something. Boys helped me up on my legs—after breakfast Fred rubbed me with camphor—felt considerably relieved. About ten AM felt about to walk. Being half way down the rapids and not trusting myself to pull at the oars, I walked down, Andy taking my place,—took the oars at the foot, supported my back. I felt all right. Came to a hell of a looking place. Here the granite comes up, two gulches having emptied their debris, forming the biggest fall we have seen. Stopped at the head, made a picture of it. Had dinner. Major examined it. Could not find a place to let down. Walls on each side vertical, smooth slippery rocks. After dinner pulled into it. In a second found our boat, the *Dean*, filled in a perfect hell of waves and foam. Got through all right. Bailed out, and watched the *Cañonita*. Only once in a while could see three heads bobbing up, but no boat could be seen. Got through all right. Bailed out and went on into more of not such large calibre. This hell hole has a fall of 60 feet in half a mile. Ran four more, then came to another monster. Concluded to make a portage—foothold being very scarce, had to maneuver very cautiously in order to get at the head of the rapid. Let one boat down to the head of the fall by line, men holding on to the craggs, wedged her in between two rocks at the head of the fall, fastened our line to the stern of the *Cañonita*, and pulled her down all right. Here we unloaded, let down by line over the swift water, passing them around the projecting rocks. Spent all afternoon to make 20 yards. Slept on the bare granite—fine bed for a lame back.

August 30

Continued all day hard at work, making a quarter of a mile. About five PM hauled the boats up for repairs. While half through with the job, here came a flood rising the river four feet in an hour and still raining but not with such a rush. Hoisted the boats up and hung them on the wall of the cañon, while we went up on the second bench to spend another night on a soft bed of hard granite.

August 31

River had fallen a little after breakfast. Got everything ready to load her as quick as possible so as not to keep her pounding too long. Everything being ready, we lowered them to the water. In lowering the *Dean*, she struck a rock which went clean through her, water filling her cabin almost instantly. Nevertheless, could not stop to repair her. Loaded the rations, which are all in rubber sacks, into her cabin, water running out on top. All was ready in a jiffy—we jumped in and soon found ourselves whirling down the tail of the rapid. Landed in a little cove for repairs, and dried out our clothes. After dinner pulled out, which was about 4 PM. Ran ten miles and ten rapids, some of them very dangerous, but shot through them like an arrow. Could not make a portage if we had wanted to. Vertical walls on both sides. Landed at the mouth of Bright Angel River at just 4:45 PM, having made the run in 45 minutes. That is what I call going it, you bet. A lone willow flourishes here—the first since Marble Cañon. Camped here on the right. River or creek is on the right.

September 1

Pulled the Major across the river. He climbed out but soon returned and started into an old roarer of a rapid three miles long. The Deacon was thrown out but hung to the side and crawled in. Lost three oars but recovered them all below the rapid in an eddy. Ran the granite down and the old red sandstone up. Ran five more rapids. Camped on the right for dinner. Made two pictures. After dinner ran the granite up again. Walls 4000 feet. Ran a big rapid. At 4 PM came to a huge fall, made a portage. Clem and myself made pictures. Climbed up for more. Camped at the foot.

September 2

Pulled out about 8 AM. Ran into a big rapid. Made it all OK. Ran three small [rapids]. Came to a fall—made a portage—ran a little way—came to another portage.

Made it, and had dinner on some rocks on the left. Pulled out after dinner, ran all afternoon, rapid after rapid, until 20 were left behind. Made 15 ½ miles, 26 rapids and two portages. Camped on the left in a gulch at the head of two rapids, which the Major says we can run, but they are busters.

September 3

Pulled out into the rapid. Made the first all right. In the second found hellish big waves, filling us instantly. Got nearly through when she was whirled about by whirlpools. Jones, in trying to hold her head down stream, overbalanced her and capsized, spilling us out. In turning over I made sure of a hold, but got fouled by an oar which knocked me six feet away from the boat, struck a whirlpool, went down head first—down, down, I went. I struggled to turn, finally succeeded. Then commenced a strange sensation. Wind by this time began to tell. I done some of the tallest kicking I ever done in my life. I thought it an age—all at once I felt myself brought up suddenly, and the next instant I had hold of the gunwale of the boat. Major and myself came up together in a boile. Must have been in the same whirlpool with me, as he spoke of being taken down by one, but he fortunately had his life preserver on or else he might have drowned, having only one arm. We soon turned her over and rowed her into an eddy, bailed out and pulled out, landed on the right side at the head of another huge monster. Had dinner, made two pictures. Pulled out into it, made it all right. A mile farther down are some old Moqui ruins on the right, and just below, a rapid two miles long. Major and Prof. examined it. Could make no portage, so had to run it. Pulled into it—while about half through, found ourselves going for the cliff on the right. Pulled hard on the left hand oar. Just missed it by six inches, or else it would have been goodbye, John. Got through all right—ran three more, then came to a young hell. In trying to land at the head on a projecting ledge of granite, Fred jumped out with the line, but missed his mark, fell backward into the river. Major sung, out, Jump boys, jump. He jumped himself to catch the line, but failed. The boat at this time was fearfully near the head of the fall, had swung round, going stern on. In passing a rock close to shore, the Deacon jumped out and just made it. At this time Fred was hanging to the line and in the water. I saw the danger, put on my life preserver, twisted the boat bow on. When I had her turned, I looked around to see where Fred was. I was overjoyed to see his head coming over the bow. In an instant he was at his oars. By this we were at the head of swift water. A huge boulder stuck out in the middle of the river. Got into its eddy. This would break at times, and lucky for us, it broke, sending a large current toward the shore. We took advantage of it, and with might and main, reached the shore. Would have gone over if a friendly rock had not stopped us near the shore. The *Cañonita* was signalled to land, who was let down by line. Camped on the left. Made 3 ¼ miles, 16 rapids, and one let down.

September 4

Commenced with a let down, then pulled out and ran rapids. Stopped on the left to make pictures, but wind blew so hard had to give it up. *Cañonita* had run down a mile. On arriving at her morrage we found dinner waiting. After dinner, pulled out into rapids. Walls about 4000 feet. Made 14 miles, 23 rapids. Camped on the left, granite down.

September 5

Pulled out at 8 AM for a little ways. Came to a let down just below. Landed on the left for pictures. Had dinner and crossed over for more pictures, then pulled out, running rapids. Came to the granite again. River immediately narrowed up. At the hea[d] of the granite found the narrowest place on the river, the river not more than 60 feet wide. Landed on the right to examine a rapid, ran it. Half mile farther down heard an angel's whisper in the roar of falling water. Could not get near to see it. Dropped to the head and found it to be a portage. Had to take the boats 200 yards up stream by line over the projecting cliff in order to get over to the left— no foothold on the right. Made the portage and pulled out. Got into the heaviest waves on the tail end of the thing—half filled. Ran more rapids. Ran the granite down and into lava or basalt, properly speaking "trap." Camped on a small bit of sand bank. Made 8 ½ miles, three portages, and ran 11 rapids.

September 6

Started out with running a huge rapid without shipping a drop of water. Ran three others—landed on the right, for pictures and a portage. Clem and myself made pictures while the rest of the boys let the boats down by line. This is the first rapid that I did not help. At the foot of the rapid a clear, cold stream [Tapeats Creek] comes in on the right—from Buckskin. Went up a ways, found several cascades—made pictures of them. Had dinner. Major went out to geologise. Started down at 3 PM. Ran two old busters. Camped at the foot of the last. Another cold stream comes in on the right—flows a large stream. Made 13–14 miles. One portage and 7 rapids run. Trap all day.

September 7

Major, Prof, Fred, Clem and myself started up the creek, but had to stop when about a mile up—it began to run in the cañon with innumerable falls. Had to climb up 800 feet. Obtained two pictures. A fine view is obtained from this height of the Grand Cañon, its walls towering up 6000 feet, and terraced back, each formation forming a terrace. Returned to camp about one PM. Had dinner and pulled out. Ran four rapids in three miles. Saw a creek on the right pouring from a cliff 200 feet high. Stopped and made two pictures. Pulled out and ran nine rapids. Some of the [illegible] ran the granite up and down in three miles. Walls are now all marble again. At 6 PM heard a shout from the mouth of a side cañon, which proved to be the Kanab Wash. Found George Adair, Joe Hamblin and [Nathan] Adams with supper for us. Rowed up the creek for half mile to a nice camp. Found one letter from Brother Dick, and lots of newspapers. One letter from Bonemorte. Mrs. Thompson had sent us down some potatoes, cheese, butter and canned fruit. She is the most thoughtful little woman I have ever known. Got in a supply of photographic materials.

September 8

Cleaned glass and packed up negatives to be sent to Kanab.

September 9

Quite a surprise this morning at breakfast. Major told us that our voyage of toil and danger was at an end on the river. Everybody felt like praising God. The party would start at noon for Kanab excepting Clem and myself to photograph up the Wash for ten miles, when horses would be sent down to us. Party left at noon. Remained here all day fixing up the traps.

September 10

Started down the Wash to the river. Made a picture looking up and one looking [down]. Came back and made a picture of the dismantled boats. Made two more, then went back to camp.

September 11

Took the traps and worked up the Wash for four miles. Walls of the cañon 3000 feet. Lower stratum yellowish looking shale—upper part marble. Returned to camp, leaving the traps behind.

September 12

Took our blankets and cooking utensils up to where we had left off. Rested, and started up. Made a picture whenever we found good light on it and sometimes would wait for light. Got up about a mile. A stratum of limestone appears on top. Shales gone under, and now marble is on the bottom walls, growing lower. Returned to our blankets and camped.

September 13

Took our blankets up to the traps—waited a long time for light—the sun only peeps down into the cañon about half an hour, but as the Cañon changes directions every quarter of a mile, we got good light often. Went up about a mile and a half. Sandstone appears on top. Walls 2000 feet high. Cañon in width about 250 feet, which is the general average. Returned to the blankets and camped.

September 14

Off with the traps. Spent a restless night. This cañon is the most gloomiest place I have ever been in—not a bird in it. The only thing of life is the bat and mosquitoes. Made some fine work today. Got up about a mile. Returned to blankets and camped.

September 15

Started up early. No rest last night. I feel lame and stiff all over. Made fine work today. Lots of cactus apples grow all along the sides of the cañon—eat lots of them every day. They are very delicious fruit and I think are very healthy. Indians live on them this season of the year. They also make wine from their juice, which they say makes drunk come. The cañon often doubles on itself, leaving only a thin wall in the bend. Photographed one today measuring in height over 2000 feet, all marble, measuring through its base only about 200 feet. Went up a mile, left the traps. Quit early for a picture, which we could not take till morning on account of light. Went back for our blankets, took them up to the shower bath, which I have discovered on a former trip down here. Here we were surprised to find Mr. Adams and Joe Hamblin, who were unpacking for tonight's camp, with ten animals for us, and the caches.

September 16

Adams and Hamblin started down for the caches on the river, while Clem and I went back to make our pictures, then returned to the shower bath. Had dinner and made a picture of it and the cañon. Left our traps and started up with our blankets to the head of the water about a mile from the bath, where we camped. Walls now about 1500 (feet). Marble, lime and sandstone.

September 17

Started back for the traps and made two pictures, when Clem unfortunately broke the slide in the plate holder. Adams and Joe Hamblin returned, having only half executed their orders concerning bringing things, only bringing part of the stuff. Camped.

September 18

Packed up this morning and started up the cañon. Made 20 miles and camped on the left near Moqui Cañon. Ran the marble, lime and red sandstone under a kind of red shale and gray sandstone up. Cañon half mile wide. No water except a pocket. Walls about 900 feet. The shales[78] are talus. The rest is vertical.

September 19

Off early, continuing our journey up the Wash. Run the shales and gray sandstone under and limestone up. Made 21 miles. Camped near a pocket under the cliff on the left.

September 20

On the road early. After riding four miles we climbed out. What a relief to the eye, after being penned up in the cañon for three months. Reached Kanab about three PM. Found Mrs. Thompson ready to welcome us. Found anactinomy [?] from Dick, also a letter from Dick['s] girls. Fred told us that we were to go to the Moqui towns. Major and Jones gone to upper Kanab.[79] Andy gone to Beaver.

September 21	Got ready to clean glass and fix up chemicals.
September 22	Went to Johnson to see Beaman.
September 23	Cleaned glass.
September 24	Clem and Fred taking barometrical observations. Prof. is taking time for longitude.
September 25	Still to work at glass.
September 26	Got trypods made.
September 27	Got a new frame for dark tent.
September 28	Got a camera box made.
September 29	Got through with glass.
September 30	Andy returned from Beaver. Mrs. Thompson, Fred and Clem went up Kanab Cañon to the Lakes, Prof. followed about noon. Party returned late at night.[80]
October 1	Fixing breachings for pack saddle.
October 2	Fixing apparatus. Major and Jones returned from their trip.[81]
October 9	Everything was ready for our Moqui trip. Took a wagon to make travel easy as far as the Colorado. Started at 10 AM, Jacob Hamblin as guide and trader. Andy, Clem and myself stopped in front of Johnson's Cañon. Took horses up four miles to water. Hamblin, who had remained behind, joined us at Johnson. Purchased several Indian trinkets, hitched up and started for Navaho Wells, where we camped. The well is simply a deep hole in the rock but always plenty of water is found in it.
October 10	About 4 AM Fred and Charley Riggs rode into camp with a dispatch for Hamblin, stating that Navajos had driven off 75 head of horses and mules from Summit Creek and had taken the southern trail. Jacob and Riggs started for the Crossing of the Fathers to head them, which can be easily done. They have to go through a narrow pass, and two men can kill a hundred. Fred after breakfast returned to Kanab,

while Andy, Clem and myself drove across the Buckskin to House Rock Spring. Arrived late at night, having come 39 miles. House Rock Valley is quite a long valley, nearly 47 miles long and from two to three miles long [wide?].

October 11

Up and off early for the Pools, where we arrived at 2 PM. J. D. Lee has quite an extensive ranch here for his stock. Stopped to water and chat with Mrs. [Rachel] Lee, one of his wives. Found the old gent, had gone to the Pario. The family here lives under a willow shelter, but a stone house is about to be erected. Strove 15 farther to a little creek where we camped. Water very alkaline. Burnt up the remains of a gold rocker belonging to some disappointed gold hunter.

October 12

Started for the mouth of the Pario where we found Jacob who had just arrived. Old John D. met us and welcomed us to his house. Had supper and returned to our old camp of last July.

October 13

Tryed to raise the *Nell* which was sunk in the river and buried in the sand. Worked two hours but without success, being wholly under water. Gave it up. Began a skiff and worked hard all day.

October 14

Still at the skiff.

October 15

Nearly finished all but pitching.

October 16

Pitched the skiff and put it in the water in the afternoon. Ferryed our traps across about 4 PM. Began to pack up and started [at] sundown. Travel till 9 PM to a creek but failed to find water. Trail very rough at the foot of the cliff.

October 17

Found water this morning in the creek that is at the head where there is a spring. After breakfast continued our journey at the foot of the cliff. The cliff now bends southward. Traveled on bank 15 miles, then found water in a gulch but so salty it would not boil beans.

October 18

Started early still going south. Nooned in a little cañon where we found water in a little pot hole, but it smelled like manure water. After dinner continued our journey. About sundown turned east. Climbed the cliff, traveled till 9 PM. Found no water for the animals. Jacob took a few canteens, knowing that water was in the neighborhood. He returned with them filled. Had supper and rolled up in blankets.

October 19

Country around is sandstone. Bare rock curiously eroded forming in good many places. Large pocket(s), found some of them filled. Water our horses. Had breakfast. Clem and myself photographed till noon then packed up and started, taking an eastern trail, leading through a valley then over a low cliff. Night overtook us. Hamblin mistook his direction and another dry camp was the result.

October 20

Up and off for water which was supposed to be found somewhere. Traveled 15 miles and found water in a large basin. Had dinner and started again. Made 15 miles more and camped near a butte. Dry camp. Met two Navajoes. *Mise am too wu* bore a recommendation from the agent of the Moquis.

October 21

Off early. After traveling five miles met a band of 9 Navajoes—Quinico, an old chief, going to the Mormon settlements to trade. Hamblin is well acquainted with Quinico and they two set up a regular pow-wow. Clem tried to trade his colt but could not get all he wanted. They offered him four blankets which I think was big. Bid the Navajoes good-bye and traveled on. Found water opposite a large white cliff at S. end in Water pockets. After quenching our thirst we went eight miles farther to the Cutchento Weep (Buffalo Land). Plenty water here—very salt. Hamblin and a party of nine passed through here in '61 on a mission to the Moquis, but were stopped by the Navajoes, then at war with all nations. Their camp was situated on a Table Rock. A mare, the property of George A. Smith, [Jr.,] ran down the trail, Smith following. On turning around a cliff, he was killed by the Navajoes. Three arrows and two bullets pierced his body. Four gray headed Navajoes agreed to pilot them back in safety, who proved to be true. George A. Smith died after surviving two hours. Navajoes scalped him and had a war dance. Party reached settlements in safety.[82]

October 22

Made a picture of the place but failed to get a good one—traps out of order. Pushed on till we reached a belt of shrub cedar where we made a dry camp.

October 23

Off for Hotoville [Hotevilla] which place we reached at noon. This place is about five miles from Oriby [Oraibi], a Moqui village. A little seep spring keeps alive a dozen little gardens 10 × 20 [ft.]. Found an old man and two squaws pulling truck. Pushed on to Oriby. Reached it a little after noon. Unpacked and were invited to eat Peakie [piki] and mellon. In the evening Jacob and myself were invited to sup. We accepted. After climbing up a ladder to the second story we were shown a place on a sheep skin. A huge earthen pot set on two stones over a small fire boiling. Directly on a huge bowl was set in front of us. It was filled up from the pot on the fire, which proved to be corn and mutton boiled into a soup. A waiter tray made of willows was next placed on the ground. An armful of cornbread resembling paper cinders rolled up was placed on the tray. Three mellons were brought on from a large stack in the corner. Everything was now ready. Three men and three women seated themselves likewise on a sheepskin. The old man gave the signal for commencement, by diving with his hand into the bowl of soup—hunted out the biggest piece of mutton. The rest followed suit. Being very hungry myself, and as the old saying is, When you are in Rome you must do as the Romans do, so I sailed in with my digits and pulled forth a dumplin'. Jacob told me not to eat, as it was prepared by the virgin of the house, who had chewed every bit before it was put into the pot. I asked what for. He told me, to arouse the animal passion of the young warior and so hasten her marriage. I allowed the dumplin to roll back, and fetched forth a leg, or rather part of a leg of mutton. I done justice to the meal. I watched the dumplin but none appeared to want it so it was left in the pot. The maiden gave a sigh. While barely through, another Mook [Moqui?] came and invited us to sup. Of course we went, but refrained from eating more soup. Ate a little peakie and mellons. Peakie is corn bread. The meal is made into a thin batter. A large slab of rock is raised about six inches from the ground—a fire built under it. As soon as hot it is poured on—no sooner on than it is done. It is rolled up in wads a foot square. There are three different kinds—red, white and blue (true American) the different kinds of corn. They build their houses on top of the cliff here in Oriby. There are three streets running parallel to each other. Houses are built of stone and clay for mortar—all joining each other, generally two stories high. A ladder is placed on the ground to the top of the first. The second story is set back, allowing a space of eight feet or more to walk on. From this you go into the second story, which has a door. A hole is made in the top of the first and a ladder placed in the hole to get down by. In summer they live in the upper story and winter in the lower. They cultivate the land, raise corn, beans, mellon, peppers and peaches. Raise lots of sheep, asses, a few horses and cattle. Men wear their hair long behind and cut even with the eyes in front. While at work they are naked excepting a breach clout. The women wear their hair long done up in a long roll hanging down on each side. Wear a black blanket dress fastened over one shoulder and a sash—that is, the marryed ones. The marriageable wear their clothes the same. Their hair is done up on the side in the shape of a ram's horn, and are as a general thing pretty—fine features.

October 24

Packed up this morning and started for "Haulpie" [Walpi] another Moqui village. After 15 miles travel reached the cliff on which were situated three towns about a quarter of a mile apart. The most southern is Haulpie standing on the end of the maissy 800 feet high. Taba [Hano][83] the most northern and the center is Suny

[Sichomovi], where we were shown to a house unoccupied. The occupant had died with the small pocks. We were greeted by two white men, one Mr. Crothers the agent's son, the other a Spanish interpreter, Mr. Wallace. They have quite a stock of Indian goods on hand, but the Indians refuse to take them. Each town wants the whole or none and so they remain stored.

October 25

Tryed photographing but the traps were out of order. Spent all day fixing them. No better result. Kept on tinkering.

October 26

Tryed our chemicals but with little better result. A high wind prevailing, gave it up and fixed a new batch.[84]

1874

September 11[85]

Left Gunnison at 3 PM. Camped late in Twelve Mile Cañon, a mile below saw-mill.

September 12

Started for Mooseneah, stopping at saw-mill to grind axes. Camped at noon at the floor of the mountain. After dinner Prof. climbed up, while Joe Hamblin and myself hunted a mule lost in the range.

September 13

Started up to the peak with the whole train. Camped right under the nipple. Prof., Joe and myself climbed to the top, taking with us theodolite and barometer. After completing observations started for camp, where we found Robert Duke waiting with a hot dinner. After dinner packed up and started down the mountain on the east side. Camped at the head of Salina Creek.

September 14

Pulled out early. Climbed out of the cañon. Found several lakes on top. Robert forgot his canteen on top, went back and got lost. Delayed us 45 minutes. Struck a trail, followed it for some time. Camped at the head of a beautiful valley for dinner. All afternoon in the valley. The trees were dressed in the most gorgeous style. Crossed a large trout stream, but could not stop. Camped near bare bush Butte on a little Creek. Saw three deer. Fired at them but they were too far.

September 15

Prof. and myself started for our peak. On arriving at the top found ourselves above the clouds. Waited two hours.

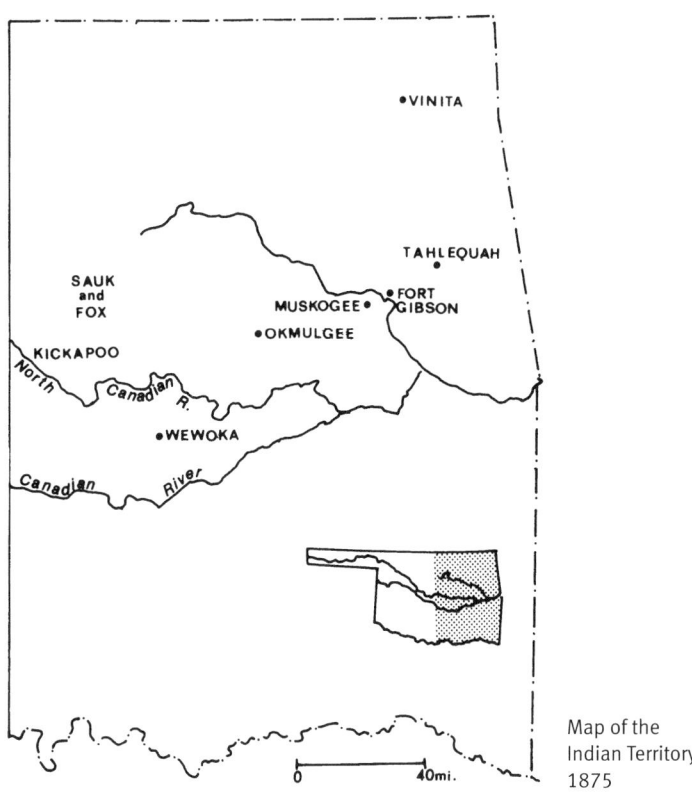

Map of the Indian Territory, 1875

May 1

S[unday]. Left Washington, D.C. for Indian Territory.[86] Breakfast at Grafton, dinner at Chilocothe, supper Cincinnati. Scenery very beautiful.

May 3

M[onday]. Arrived at East St. Louis this morning. Crossed Mississippi on suspension bridge. Stopped at the Planters [Hotel]. Received telegram from Major [G.W.] Ingalls.[87]

May 4

T[uesday]. Took a walk through town. Full of business. Mailed a postal order to Mahuken [?] Ingalls arrived in town at noon, his brother and a Mr. Cobble at night.

May 5

W[ednesday]. Took a Pullman for Muskogee. Ingalls took out 40,000 Dollars for the natives. Splendid country for stock-raising. Sceneray very fine up the Missouri.

Changed trains at Vinita on the K. M. and T. Road [Missouri, Kansas and Texas Railroad]. Major [Renfrew M.] Roberts met us at this point.

May 6 — Th[ursday]. Arrived at Muskogee. Small town, a dozen houses, a few stores, rest nigger shanties. I was struck with the intimacy of Negroes and Indians. On inquiry found they intermarry. Fixed up my chemicals.

May 7 — F[riday]. Ingalls, Roberts and Doc. [Myron P.] Roberts left for Okmulkee 50 miles from here. My traps being too heavy had to get another wagon, which arrived about 4 PM when I left town.

May 8 — S[aturday]. Arrived at Okmulkee about 7 PM. Found Ingalls, Roberts and Gen. [John Peter Cleaver] Shank[s].[88]

May 9 — S[unday]. Fixed my traps and made a picture of Lievers building.

May 10 — M[onday]. Photographed Cheyenne Indians.

May 11 — T[uesday]. Making pictures of Indians.

May 12 — W[ednesday]. Making pictures of Indians.

May 13 — Th[ursday]. Pho[tographed] the council and Tocopes.[89]

May 14 — F[riday]. Phot[ographed] the Pawnees.

May 15 — Sat[urday]. Left for Weewaukee [Wewoka].[90] Camped at Ike Smith's, 15 miles from Okmulkee. Mr. Brown, a half breed Seminole accompanied me.[91]

May 16 — S[unday]. Started early. Crossed many prairies. Fine stock country. Arrived at Wewaukee about 7 PM. Stopped six miles east of Wewaukee and made two pictures of cascades. Very beautiful. Found the river up and booming. Crossed on a large sycamore which had been felled across the stream. Stopped at Mrs. Lilly, a widow who keeps a hash house, and is assisted by her daughter and sister.

May 17 — Made pictures of council house and church.

May 18 — T[uesday]. T. Ingalls and Major Roberts left this morning, also General Shanks, with the Cheyenne Indians in charge of Phil McCusker, interpreter for the Wichita Agency.[92] Made a picture of John Chupco,[93] Factors church and a family group of Mr. Brown's.

May 19   W[ednesday]. Started for Col[onel John] Jumper's, Chief of Seminoles,[94] 15 miles. Made a picture of his church. Started back about a mile to the trader, a Mr. Brown, half breed, whose father had been an English nobleman. I was show[n] all his valuables which are still preserved. He was or had been Prof. of Edenbourgh [Edinburgh] College.

May 20   T[hursday]. Started back for Weewaukee. Made a picture of Brown's house then came 15 miles and camped at a Mr. [E. A.] Eggleston, a rancher and trader. A storm came up, the worst I have ever seen. It lasted for ten hours.

May 21   F[riday]. River high. Could not cross Little Wewaukee. Took a ride downstream. Came to a house, found a pair of twins spread out on a blanket, naked in the shade. They looked well, and judging from the mother's fulness they were supplied with plenty of clabber. What would I not have given for a photograph of all, but I had not the collodion to do it with.

May 22   S[aturday]. Started off this morning, the creek being fordable. We went five miles then to be stopped by the high water in the North Fork of the Canadian. Camped with an Indian who was a poligamist, having two wives. He treated us first class. Dr. Robert, correspondent for the Inter Ocean, and a nigger, a lazy brute.

May 23   S[unday]. Started across the North Fork of the Canadian River. A lot of Indians crossed the same time going to camp meeting. Arrived at Okmulkee at 6 PM.

May 24   M[onday]. Started for Muskogee. Joe Cornells here took his own team. Had a lady passenger weighing about 250, a half breed at that. Camp 25 miles out.

May 25   T[uesday]. Arrived at Muskogee about 11 AM.

May 26   W[ednesday]. Started for Tahlequahm [Tahlequah], the capitol of the Cherokee Nation. Arrived at 7 PM. Stopped at the Rev. J[ohn] B[uttrick] Jones.[95]

May 27   T[hursday]. Made pictures of house and family and three young half breeds.

May 28	F[riday]. Went to Hildebrand Falls, seven miles from T[ahlequah]. Mr. Stevens and Williams accompanied Gus and myself. Made the pictures and returned by 7 PM.
May 29	S[aturday]. Started for the Ladies Seminary[96] with same party. Made pictures of all the pupils. Quite a lot of pretty girls. Started for the Orphand Asylum, but too late to make a picture, as the sun was in the camera. Returned to town and made picture of council house.[97] Returned to Jones'.
May 30	Sunday. Fasted.
May 31	M[onday]. Returned to Ft. Gibson. Made picture of Asylum and one cascade.
June 1	T[uesday]. Made picture of Chief Ross' house and one of himself.[98] Returned to Muscogee.
June 2	W[ednesday]. Made arrangement to go with Major [John H.] Pickering to Sac and Fox Agency.[99]
June 3	Th[ursday]. Started this morning—two wagons. Pickering, wife and adopted child, and a Prof. Pickett, clerk for Pickering. Mr. Whisler [John Whistler], a rancher,[100] took my traps. Gus and myself reached Okmulkee about 7 PM. Had a good time with Mr. Austin,[101] the ever-attentive bookkeeper for Mr. Lievers. After lubricating from a flask of brandy we had supper, on which occasion Billy was quite flowery. Topic, calling of man.
June 4	F[riday]. Started this morning. A storm came on while a few miles out from Okmulkee. Stopped for dinner about 15 miles out. About four PM came to a creek which was up and booming, which prevented the wagons from crossing. The Pickerings crossed on a foot log and started for a ranch, mile and a half farther on. Whisler, Gus and myself remained. Storm came on about midnight. The creek, which had gone down by 11 PM now raised again.
June 5	S[aturday]. Had to wait till 2 PM, when we crossed. Water touched the wagon box and a little went in and wet my glass, damn it. An hour's drive brought us to deep fork. Here we went across on a foot log in search of Pickering who had come down to the crossing. We unloaded our wagon and brought it to the opposite side on the foot log, then attached a line to the wagon tongue, which an Indian brought across by swimming. Camped at night at a Mr. Butler, a Negro-Indian.

June 6	S[unday]. Started this morning early. Raining nearly all day. Arrived at the Agency about 3 PM.
June 7	M[onday]. Fixed my traps and started for the Mission House to make a picture but traps would not work, so returned and fixed up for tomorrow.
June 8	T[uesday]. Wind blowing like hell. Made two large and three stereos of Agency and Mission House.
June 9	W[ednesday]. The Indians held a council. Photographed 3 women and two men. Stereo, out of fix.
June 10	Th[ursday]. Started this morning for Kickapoo towns with Gus and the interpreter. When seven miles from towns broke wagon tongue, which was spliced at the ration house. Quite a number here drawing rations.

1875: A Letter

Dear Brother:

While I have time I will spin you my log. On my arrival at Muskogee I was greeted by a sight little expected—a big burly Indian kissing a Negroe wench. On inquiring found that the Indians and niggers intermarry. This is just the contrary among the wild tribes. They hate a nigger and think him a being of the devil, an Unupits, but only the Creek Indians believe in the amalgamation of red and black. In the Cherokee where they are nearly all half breeds it is forbid [den] by law and a prison offence. I spent the day at Muskogee, and on the following day set out with an old darky for Okmulgee, 80 miles inland. Major Ingalls had preceded me in his carriage. About 7 PM my King of Sheba arrived for my utencils. His team consisted of two mules whose sides were fastened to a single straight gut without a stomach, and an old fashioned home-made wagon. At the sight of these animals led me to ask how soon he expected to make Okmulgee. He assured me that he would land me at that place tomorrow's sundown. Off we started. I found the old man very inquisitive, but like the majority of Negros, very ignorant, and believed in the Baptist doctrines, but I think the old man was terrible puzzled concerning his future state. After we arrived at Okmulgee on the time promised by him, I bid him goodbye. Here I found Major Ingalls arranging matters. The Grand Council was in session and all Nations were represented living in this territory. I grouped the whole and made a No. 1 negative. Here I found six Cheyennes who had just left the war path, all strappen big fellows. I took them among the rocks and set them up as food for my camera. I stripped them to the buff, not a stitch on them except a breach clout and succeeded in making pictures of them all. I spent just a week here packing up odds and ends. From here I went to Weewaukee, Seminole country. You see, Dick, everything ends with a *kee*, but damn it, no whiskey. The half breeds and bloods find a substitute in "Jamaica ginger", "Pain Killer", Mustang liniment, and "Railways' Ready Relief," and a little of it converts them into the once savage "plume taker." I engaged a man to take me over, but owing to sickness he was prevented from going, so he let me have his mules, two good animals, and a spring wagon. Council had adjourned and Mr. Brown, a half breed member of the council for the Seminoles, returned home with me. On our way found two little cascades, which I photographed. After a day and a half's journey we arrived at the village, consisting of half dozen houses or more, blacksmith shop, saw and grist mill, the latter owned and run by a white trader named Long. On arriving found Ingalls, General Shanks, and Major Roberts had arrived. Stopped at the house of Mrs. Lilly, a widow. She runs a hash house, assisted by her sister and a daughter, a beautiful Lilly of sweet 20 or thereabouts. I got along nicely with the young miss, wheather it was to make herself amiable, or simply to work herself into my good graces for a picture of herself, I was in doubt. However, I pleased her. She made some attempt at singing and playing an accordion, but neither was first class. After staying a few days here I went to the home of that renown[ed] chieftain, old John Jumper, who ten years ago counted his scalps by the hundred, and who then wore a buckskin shirt trimmed with the hair of his victims, but now a peaceable farmer and expounder of the Gospel. Liken to Paul, he had seen a vision, buried the tomahawk and drove the scalping knife into a large sycamore. Arriving at his home found him tending to his children. He formerly was a poligamist and had three wives, but when he joined the church he was told to let two of them go and retain only one. He kept the children. What became of the women I could not learn. Close by his house he has erected a large building 10×60 [ft.] which is used as a school house and church. Quite a bell is suspended in the cupelo on top. A half breed Choctaw lady is the teacher of the young ideas. Now, Dick, I must describe to you this really beautiful country. Ride with me in mind: The road to this Col. Jumper's place from Weewaukee winds through low foothills through which a hundred little streamlets wind their way to the Canadian River. Their waters are clear and sparkling, and as they tumble over some moss-covered ledge it is dashed into foam. The banks are fringed with foliage equal to the tropics. Here the stately oak and pecan tree, with their rich garb of bright green, luxuriate, and their branches form a lattice work for the wild grape, honeysuckle and ivey. The prairie is covered with wild flowers of every hue and the whole air is perfumed with [their] sweet balm, especially in the morning, and a million birds greet the sweet dawn of day. And at midnight, if camped in some beautiful grove by some trinkling spring, your dreams are disturbed by the sweetest singing of the nightingale. What a dreamland this is. It is a garden spot of America. But soon I must exchange this for the sage brush, sandy plains, and alkali water. I, on my return to Muskogee, took a trip into the Cherokee Nation. I went to Tahlequah, the capitol of the Nation. Mostly half breeds reside here and some are so white that you cannot tell them from the bona fide white. The young misses are exceedingly beautiful. They retain the fascinating black eyes and glossy black hair of the Indian. I went to the Ladies' Seminary to photograph this piece of Indian enterprise. Being well backed by letters of introduction from prominent men of Tahlequah, I started for the Institute, which is situated about three miles south of the town. On my arrival, was kindly cared for by the principal teacher, a young miss on the yellow-leaf side of thirty, who introduced Prof. Hillers, Govnt. Photo, to her young and beautiful half, quarter, and octavo blood Indians. I soon made known my wants, stating that I desired a picture of their buildings and their beautiful selves, for to be put on exhibition at the Centennial. Soon I saw them arrayed in their finest, and really, their dresses would not be sneezed at in any metropolis. A very beautiful quarter-blood came to me after my grouping had been completed, and with the sweet bewitching smiles, asked me to place her in position. Of course, I would, and show her beautiful form such as but few women possess, to an advantage. At the sight of her large lustrous black eyes and long jet black hair and her regular cut features, and a smile so captivating, threw me almost into cupid's arms, and for a moment I forgot my bachelor's resolution. I placed her near the center. A few small stones were lying loose on the ground. I told her to be seated, three-quarters front. I placed her delicate little No. 2 slipper foot on a small stone, her head resting lightly on three fingers, her eyes turned up and a sweet smile spread all over her face. Her dress of white and a black lace thrown over her shoulders, she set like a queen. I made the picture which was truly a good one, and how it happened I could not tell, for my nerves were all unstrung, my hand trembling, I could scarcely handle the plate. After they had all seen it and admired it, they returned to their rooms. What had become of my queen? I could see her nowheres. I felt sick. I packed up, took my picture into the kitchen to varnish it by the fire. Another crowd came rushing in who wanted one more look. Of course I gratified their desire. And I wished my Ideal would come in and have a look—not that I cared to have her look, but I wanted to feast my own eyes. I was doomed to disappointment. I bid them goodbye, mounted my horse and sadly rode away. The landscape so beautiful about the Seminary, had no charm for me—my thoughts were of the beautiful Indian maid. I asked myself if this could be anything like love.

Well, I concluded, if this painful desire was called love, I wanted to be rid of it, and that very soon. I changed my thoughts to the days when I was tempest-tossed on the Colorado River, shooting over falls and rapids, climbing pinnacles and towers, and rushing rock-ribbed gulches—a life full of adventure. While in this revery I heard the sound of a horse's hoofs and the breaking of limbs. Involuntarily my hand went to my pistol holster, pulled it out and looked to the priming and before I could replace it a sweet voice entreated me not to shoot her. If, no doubt, could I have seen my face I would have been surprised at the many colors. Mounted on a thoroughbred sat my Vision. The color had deepened on her face from the exercise and looked more beautiful than ever. She begged pardon for her intrusion, but she thought of going home and as my road was nearly in the same direction she thought of placing herself under my protection (thanks for your good opinion). We rode about two miles when my road led in an opposite direction. She told of a cascade which was situated near her home, which induced me to change my mind. As it was early I could have plenty of time to return before night. So I stated to her that I would go with her and see the falls. She seemed pleased. We talked of everything, about their beautiful country, people and music. We arrived at her father's house early afternoon. I unpacked, while she oppened the bars to allow my animals to go into pasture. I asked her motive. She responded that my animals would be safe in there for the night. I said No, I must go back to the city. All right, we will see about that. She called a helper to have my blankets and traps stored away until wanted. Just imagine your poor old brother for a moment, dressed in a gray shirt with large collar, black silk sailor's scarf tied with studied negligence, broad sombrero, a pair of 5 dollar cowhide boots which were adorned at the heels by a pair of Mexican silver spurs, pants inside, full grown beard and long hair. In this style my maid led me to the house, which set back in a yard full of flowers. The first we met was the Colonel, her father whom she introduced me to as a Jerome. The old gent was pleased to have someone with whom he could talk. The old lady was a pleasant old lady, rather fat and a half breed. The old man talked politics, and a rebel sympathiser, an argument which I studiously avoided. A lunch had been set out for my Alvoretta and myself. After a thorough absolution from dirt I sat down and enjoyed a first class lunch. She had read Scribner's Magazines and had seen my pictures in them,[102] which of course she praised to the skys, and no doubt she meant. I proposed a walk to the falls, which of course the Colonel would not listen to, not today—tomorrow. But, my dear sir, I must go back to the city. But don't mention city—you're my guest for the night and the other nights while you stay in the Nation. Just make yourself at home. Here, Al, play some music for the Professor, to get him out of the notion of going to town. I was very easily persuaded.

The house was elegantly furnished, the parlor was superb, an 800 dollar piano graced one side of it. I was thunderstruck when my vision began "Ever of thee I (am) fondly dreaming." I think my hands trembled while I turned the leaves for her. Her sweet music and song caryed me back to days gone by when I was courting a young widow down south in Georgia, but such is life (!). She begged me to sing, which I did and then sang a duet, "What are the wild waves saying," she was so pleased that we sang it twice over. The old man was a great smoker and seemed to enjoy to make smoke rings while so enjoying ourselves on the varanda, the Colonel's chosen place for the enjoyment of his Havanna. I heard the quavering chords of a guitar in the garden below playing, "Come love, come, ere the night torches pale," etc. As the last sweet note had died away the Colonel proposed to retire, which I also was pleased to do. As I slept long this morning, something very unusual with me, found my host waiting breakfast for me, a beautiful boquet of flowers lay by the side of my plate. I thanked the unknown donor. Breakfast over, we started for the falls. The Colonel feeling somewhat indisposed, excused himself, so I had my queen for a companion and guide. The road to the cascade let through an oak grove.[103]

Second Powell River Expedition, May 22, 1871, at Green River Station, Wyoming Territory.
Left boat (l–r), E. O. Beaman, Andrew Hattan, Walter Clement Powell;
center boat (l–r), Stephen Vandiver Jones, John K. Hillers, John Wesley Powell,
Frederick S. Dellenbaugh; right boat (l–r), Almon Harris Thompson, John F. Steward,
Francis Marion Bishop. Photographer unknown.

Ashley's Falls, Green River. Note boat at left. E. O. Beaman photograph.

Harrel's Party, Brown's Park, Green River, 1871. E. O. Beaman photograph.

Powell expedition members at Dodd's cabin, confluence of Uintah and Green Rivers, 1871.
1. F. S. Dellenbaugh, 2. J. K. Hillers, 3. A. H. Thompson, 4. E. O. Beaman (?), 5. W. C. Powell,
6. J. F. Steward, 7. A. J. Hattan, 8. F. M. Bishop. Photographer unknown.

Boats in Desolation Canyon, 1871. Probably J. F. Steward seated in "Nellie Powell."
Note boat being "lined" through rapids at left. E. O. Beaman photograph.

"Moqui" ruin on ledge, confluence of White Canyon and Glen Canyon, 1872.
J. Fennemore and J. K. Hillers photograph.

Three Patriarchs, Zion Canyon, Utah, 1872.

Zion Canyon, Utah, 1872.

Kolob Plateau, Utah, 1872.

Inner Gorge of Grand Canyon, Arizona, August 1873. Note man on cliff.

Grand Canyon looking north, 1872. J. Fennemore and J. Hillers photograph.

Marble Canyon, Arizona, 1872.

Marble Pinnacle, Kanab Canyon, Arizona, 1872.

"Summer Home Under a Cedar Tree." Chuarumpeak, leader of the Kaibab Paiute band
is second from the left. Near Kanab, Utah, 1873.

"Mother and Child." Kaibab Paiute near Kanab, Utah, 1872(?).
Note the museum accession number and the word "Colorado"
on the bodice of the woman's dress.

"The Tavokoki or Circle Dance." Kaibab Paiute near Kanab, Utah, 1872(?).

Three Paiute women in native dress. Near Las Vegas, Nevada, 1873.

An 1872 tribal council near Kanab, Utah. John Wesley Powell nearest to the camera
on the right; Jacob Hamblin seated next to him.

"Chuarumpeak Shooting a Rabbit." Kaibab Paiute men near Kanab, Utah, 1873.

"Ta-noats." Las Vegas Paiute. Las Vegas, Nevada, 1873.

"E-nu-ints-i-gaip, an old man." Las Vegas, Nevada, 1873.

A Paiute woman grinding seed in 1872 near Kanab, Utah.

J. K. Hillers sewing. Red Canyon Park, Green River, 1874.

J. K. Hillers, probably near Kanab, Utah, 1872 or 1873.
Photographer unknown, possibly J. Fennemore.

High Falls, Bullion Canyon, Utah, 1874.

Pilling's Cascade, Bullion Canyon, Utah, 1874.

"The Mirror Case." John Wesley Powell and Uintah Indian Woman,
Uintah Valley, Utah, 1874(?).

"Sai-ar and His Family." Uintah Valley, Utah, 1874.

"A Domestic Camp Scene." Uintah Valley, Utah, 1874.

A Paiute encampment on the Rio Virgin, a tributary of the Colorado, near St. George, Utah.

Repairing boats in the Grand Canyon, 1872.

Just inside the entrance to Canyon of Ladore, 1874. J. C. Pilling is shown in photograph.

Powell expedition camp along the Green River, 1871. E. O. Beaman photograph.

Bow Knot Bend from the Green River, 1871. E. O. Beaman photograph.

Bow Knot Bend from on top, Green River, 1871. E. O. Beaman photograph.

Side canyon, Glen Canyon, August 1872. J. Fennemore and J. K. Hillers photograph.

J. K. Hillers working with his photographic equipment, Aquarius Plateau, Utah, 1872.
J. Fennemore photograph.

Epilogue

The photographs Hillers took in Indian Territory in 1875 were used in the Smithsonian Institution exhibit at the 1876 Philadelphia Centennial Exhibition.

In later years Powell sent Hillers on several expeditions to the West for the Bureau of Ethnology and the Geological Survey. One of the more important of these was an expedition to the pueblos of New Mexico and Arizona and to Canyon de Chelly, Arizona, in 1879. Hillers was accompanied by James Stevenson and Frank Hamilton Cushing.[104]

Hillers was chief photographer of the Geological Survey until 1900. He continued working for the Survey on a part time basis until 1919 when ill health and advancing age forced him to retire.

Hillers's career as a professional photographer spanned a period of nearly fifty years. It began through a chance meeting of the young teamster Hillers and the young scientist Powell in 1871 in Salt Lake City. Through his research and administration of research Powell came to have an enormous influence on the courses of American anthropology, geology, and conservation practices.[105] Hillers, through his photography, played a vital role in documenting the researches of Powell and his associates. On another level Hillers's photographs remain a monument to his skill and ability. He was truly one of the great photographers of the nineteenth century American West.[106]

Notes to the Introduction

1. Robert Taft, *Photography and the American Scene: A Social History, 1839–1889* (New York, 1964); Maurice Frink with Casey Barthelmess, *Photographer on an Army Mule* (Norman, Okla., 1965); James D. Horan, *Timothy O'Sullivan: America's Forgotten Photographer* (New York, 1966); William H. Goetzmann, "Images of Progress, the Camera Becomes Part of Western Exploration," in William H. Goetzmann, *Exploration and Empire: The Explorer and the Scientist in the Winning of the American West* (New York, 1966), pp. 603–48.

2. Goetzmann, *Exploration and Empire*, pp. 231–331; William H. Goetzmann, *Army Exploration in the American West* (New Haven, 1959).

3. Goetzmann, *Exploration and Empire*; Richard A. Bartlett, *Great Surveys of the American West* (Norman, Okla., 1962).

4. William C. Darrah, *Powell of the Colorado* (Princeton, 1951); Wallace Stegner, *Beyond the Hundredth Meridian: John Wesley Powell and the Second Opening of the West* (Boston, 1954); Donald Worster, *A River Running West: The Life of John Wesley Powell* (New York and Oxford, 2001).

5. Elmo Scott Watson, ed., *The Professor Goes West: Illinois Wesleyan University—Reports of Major John Wesley Powell's Explorations, 1867–1874* (Bloomington, Illinois, 1954), 111.

6. Ibid., p. 24.

7. Don D. Fowler, Foreword, John Wesley Powell, *Down the Colorado: Diary of the First Trip Through the Grand Canyon* (New York, 1969), p. 14.

8. E.g., the illustrations in Clarence E. Dutton, *Report on the Geology of the High Plateaus of Utah* (Washington, 1880), and in John Wesley Powell, *Exploration of the Colorado River of the West and Its Tributaries* (Washington, 1875).

9. Don D. Fowler and Catherine S. Fowler, eds., "Anthropology of the Numa: John Wesley Powell's Manuscripts on the Numic Peoples of Western North America, 1868–1880," *Smithsonian Contributions to Anthropology*, vol. 14 (1971).

10. John Wesley Powell and G. W. Ingalls, *Report of Special Commissioners J. W. Powell and G. W. Ingalls on the Condition of the Ute Indians of Utah . . .* (Washington, 1874).

11. See Julian H. Steward, *. . . Notes on Hillers' Photographs of the Paiute and Ute Indians Taken on the Powell Expedition of 1873*, Smithsonian Miscellaneous Collection, vol. 98, no. 18 (Washington, 1939), and Robert C. Euler, *Southern Paiute Ethnohistory*, University of Utah Anthropological Papers no. 78 (Glen Canyon Series no. 28) (Salt Lake City, 1966), Appendix I.

12. Taft, *Photography and the American Scene*, pp. 17–21.

13. Darrah, *Powell of the Colorado*, p. 182, n. 7. An itemized statement in the archives of the Office of the Secretary in the Smithsonian Institution lists the sum of $2,413.89 for photographic materials and the employment of Beaman in 1872. The statement includes $1,173.70 for equipment and supplies, $800.00 to Beaman "for services rendered," and $440.19 to W. H. Jackson. A separate itemized bill from Jackson, dated July 5, 1872, lists charges for printing 4,288 stereographs and for mounting 549 others between March 10 and July 5, 1872. ("Statement of Expenditures for Photographic Apparatus and

Material, *and* also for the Employment of a Photographic Artist, for the Exploring Expedition of the Colorado River of the West," J. W. Powell to Joseph Henry, December 12, 1872, Henry Papers, Archives, Office of the Secretary, Smithsonian Institution.)

In addition to selling them, Powell used stereographs in other ways. During appropriation hearings and at other critical times in Congress, Powell sent sets of views to various key congressmen. In 1877 Powell was particularly anxious about the continuation of his appropriation. The Rocky Mountain Survey letter books show that between January and March of 1877, sets of views were sent to thirteen congressmen and other politically powerful men including Simon B. Cameron, a senator from Pennsylvania and previously Lincoln's secretary of war, and John Sherman, senator from Ohio and secretary of the treasury under Hayes.

During the period 1871–75 several sets of views were made, some scenic, some of Indians. Sets include "Views of the Green River," "Views on the Colorado River," "Views on the Sevier River," each with sub-series, and sets of views of Indians of the Colorado River, the Navajo and the Hopi and Zuni. A photographic catalog (in the Smithsonian National Anthropological Archives), the "Catalog of Negatives, River, Land and Ethnographic, 1871–1876," lists 131 negatives by Beaman, 71 by Fennemore, 27 by Fennemore and Hillers, and 368 by Hillers. Sets of the views in the Denver Public Library indicate that some were published by J. F. Jarvis (who for a time worked for W. H. Jackson), in Washington, D.C., and others by the William B. Holmes Company in New York City. Some views were copyrighted by Powell in 1874. Others, issued after 1879, have the statement, "Smithsonian Institution, Bureau of Ethnology, compliments of J. W. Powell," printed on the back. For additional information on the provenance of Hillers's photographs, see Don D. Fowler, Appendix, A Note in Attribution, *The Western Photographs of John K. Hillers: Myself in the Water* (Washington D.C., 1989), pp. 158–61.

14. Darrah, *Powell of the Colorado*, p. 213.

15. Ibid., pp. 254 ff.

16. John Wesley Powell, "First Annual Report of the Director of the Bureau of Ethnology (for the fiscal year 1879–80)," *Bureau of Ethnology, First Annual Report* (Washington, 1881), p. xxx; Raymond S. Brandes, *Frank Hamilton Cushing: Pioneer Americanist*, Ph.D. diss., University of Arizona, Tucson (University Microfilms 65-9951), pp. 25–35.

17. United States Civil Service Commission, *Official Register of the United States, Directory, 1816–1959* (Washington, 1907).

18. This brief biographical sketch is adapted from an obituary in the Washington, D.C. *Star*, November 16, 1925; from William C. Darrah, "Beaman, Fennemore, Hillers, Dellenbaugh, Johnson, and Hattan," *Utah Historical Quarterly*, vols. 16–17 (1948–49):

495–97; and from personal interviews with Mrs. J. K. Hillers, Jr., of Washington, D.C.

19. Darrah, *Powell of the Colorado*, p. 212.

20. F. S. Dellenbaugh to Robert Taft, November 19, 1932, printed in Taft, *Photography and the American Scene*, pp. 289–91.

21. John Gregory Bourke, *The Snake-dance of the Moquis of Arizona, Being a Narrative of a Journey from Santa Fé, New Mexico, to the Villages of the Moqui Indians of Arizona . . .* (New York, 1884), p. 354.

22. Civil Service Commission, *Official Register*, p. 78.

23. Darrah, *Powell of the Colorado*, p. 397.

24. Stegner, *Beyond the Hundredth Meridian*, p. 268. Don D. Fowler, *The Glen Canyon Country: A Personal Memoir* (Salt Lake City, 2011), p. 90.

25. The journals of George Y. Bradley, John C. Sumner, and John Wesley Powell, some letters of O. G. Howland, and other related materials from the 1869 river expedition (edited by W. G. Darrah) are found in *Utah Historical Quarterly*, vol. 15 (1947): 1–148. See also O. Dock Marston, "The Lost Journal of John Colton Sumner," *Utah Historical Quarterly*, vol. 37 (1969): 173–89. See also Michael Ghiglieri, *First Through the Grand Canyon: The Secret Journals and Letters of the 1869 Crew Who Explored the Green and Colorado Rivers* (Flagstaff, 2003).

The documentation of the second river trip of 1871–72 is extensive. It includes the journal and letters of Francis Marion Bishop, ed. Charles Kelly, *Utah Historical Quarterly*, vol. 15 (1947): 154–253; the journal of Stephen Vandiver Jones, ed. Herbert E. Gregory, *Utah Historical Quarterly*, vols. 16–17 (1948–49): 19–174; the journal and newspaper articles written by Walter Clement Powell, ed. Charles Kelly, *Utah Historical Quarterly*, vols. 16–17 (1948–49): 257–478; the diary of Almon Harris Thompson, ed. Herbert E. Gregory, *Utah Historical Quarterly*, vol. 7 (1939): 1–140; John Wesley Powell, "John Wesley Powell's Journal: Colorado River Exploration, 1871–1872," ed. Don D. Fowler and C. S. Fowler, *Smithsonian Journal of History*, vol. 3, no. 2 (1968), pp. 1–44; Frederick S. Dellenbaugh, *A Canyon Voyage: The Narrative of the Second Powell Expedition Down the Green-Colorado River from Wyoming, and the Explorations on Land, in the Years 1871 and 1872* (1908, reprint ed., New Haven and London, 1926); and E. O. Beaman, "The Cañon of the Colorado, and the Moquis Pueblos," *Appleton's Journal*, vol. 11 (1874): 481–84, 513–16, 545–48, 590–93, 623–26, 641–44, 686–89.

Powell's own published version of the two river trips, written as if all events took place in 1869, is contained in J. W. Powell, *Exploration of the Colorado River of the West*. Other related materials are found in Watson, *Professor Goes West*; Paul Meadows, "John Wesley Powell: Frontiersman of Science," *University of Nebraska Studies*, n.s., 10 (1952): 1–35; and Lindsey

G. Morris, "John Wesley Powell: Scientist and Educator," *Illinois State University Journal*, vol. 31, no. 3 (1969), pp. 3–46.

26. Stephen Vandiver Jones, "Journal of Stephen Vandiver Jones," ed. Herbert E. Gregory, *Utah Historical Quarterly*, vols. 16–17 (1948–49): 57–58.

Notes to Hillers's Journal

1. According to Stephen Vandiver Jones ("Journal of Stephen Vandiver Jones," ed. Herbert E. Gregory, *Utah Historical Quarterly*, vols. 16–17 [1948–49]: 24), the cabin was owned by "a white man who keeps the ferry at the [Green River] station."

2. The party was moving into Horseshoe Canyon.

3. "Today with Bismark [Hillers] I climbed mountain S[outh] of Kingfisher Greek" (John Wesley Powell, "John Wesley Powell's Journal: Colorado River Exploration, 1871–1872," ed. Don D. Fowler and C. S. Fowler, *Smithsonian Journal of History*, vol. 3, no. 2 [1968], p. 9).

4. See George Young Bradley, "George Y. Bradley's Journal," ed. William C. Darrah, *Utah Historical Quarterly*, vol. 15 (1947): 33–34.

5. "One rapid where Theodore Hook…was drowned in 1869…gave us no trouble" (Frederick S. Dellenbaugh, *A Canyon Voyage: The Narrative of the Second Powell Expedition Down the Green-Colorado River from Wyoming, and the Explorations on Land, in the Years 1871 and 1872* [1908, reprint ed., New Haven and London, 1926], p. 25). Dellenbaugh notes that Hook was buried nearby; cf. Almon Harris Thompson, "Diary of Almon Harris Thompson," ed. Herbert E. Gregory, *Utah Historical Quarterly*, vol. 7 (1939): 15.

6. Now called Trail Creek and Allen Creek respectively (Jones, "Journal," p. 31, n. 13).

7. Lena was Bishop's sister (Francis Marion Bishop, "Capt. Francis Marion Bishop's Journal," ed. Charles Kelly, *Utah Historical Quarterly*, vol. 15 [1947]: 170).

8. Interestingly, Powell ("Colorado River Exploration, 1871–1872," p. 13) makes no note of Richardson leaving the party, though other members do (Bishop, "Journal," p. 170; Jones, "Journal," p. 35).

9. The name "Lodore" was apparently suggested by Robert Southey's poem, "The Cataract of Lodore," which Powell evidently knew by heart.

10. Named for J. F. Steward's daughter (Walter Clement Powell, "Journal of W. C. Powell," ed. Charles Kelly, *Utah Historical Quarterly*, vols. 16–17 [1948–49]: 273; Dellenbaugh, *A Canyon Voyage*, p. 35).

11. Named by J. W. Powell on the second trip for the constellation Lyra (J. W. Powell, "Colorado River Exploration, 1871–1872," p. 15).

12. "Named for William H. Dunn, who, with the Howland brothers, left the 1869 river party in the Grand Canyon, climbed out onto the north rim, and was slain by Shivwits Indians (see William C. Darrah, *Powell of the Colorado* [Princeton, 1951], pp. 140–41).

13. Thompson, "Diary," p. 20, calls this "Serratus Ridge."

14. Powell's men called this area Echo Park (Jones, "Journal," p. 42; Dellenbaugh, *A Canyon Voyage*, p. 49); it is also called Pat's Hole.

15. J. W. Powell later renamed this creek "Bishop's Creek" for F. M. Bishop; it is now called "Jones Hole Creek" for S. V. Jones (Jones, "Journal," p. 45, n. 24).

16. Hillers's diary is the only extant description of the hike made by him and J. W. Powell to the Uintah Indian agency. J. W. Powell's diary ("Colorado River Exploration, 1871–1872") stops on July 7, 1871, and resumes on September 2, 1871, when he rejoined the river party at Gunnison's Crossing.

17. Jones, "Journal," p. 49, calls this man "To-a-quan-av."

18. This man was usually known as "Tabby" (see K. B. Garter, comp., *Indian Chiefs of Pioneer Days*, 2nd ed. [Salt Lake City, 1932]).

19. Following the incident, "'Bismark' and I concluded to move our bed within doors and slept for the first and only night since May 20th under a roof" (Jones, "Journal," p. 50).

20. Also called Gray Canyon.

21. James White, a trapper, appeared at Callville (established in 1864 by Anson Call at a point on the Colorado River just above the present site of Hoover Dam) below the Grand Canyon on September 8, 1867, claiming to have escaped from Indians, built a raft, and floated down through the Grand Canyon for eleven days. He was starving, badly sunburned, and demented. His story was printed in January 1869, in the *Rocky Mountain News*. There is some evidence that Powell sought him out and questioned him about the trip in early 1869 (Wallace Stegner, *Beyond the Hundredth Meridian: John Wesley Powell and the Second Opening of the West* [Boston, 1954], pp. 33, 376). Robert Brewster Stanton in his *Colorado River Controversies* (New York, 1932), pp. 70–93, investigated the story at length and concluded that White had perhaps floated from the Grand Wash Cliffs, below Grand Canyon, to Callville.

22. This ruin stood on a prominent point overlooking the mouth of White Canyon. It is described in Ted Weller, "San Juan Triangle Survey," in *The Glen Canyon Archeological Survey*, University of Utah Anthropological Papers no. 39, pt. 2 (Glen Canyon Series no. 6) (Salt Lake City, 1959), pp. 543–669.

23. This ruin was excavated in 1958–59 by the University of Utah (William D. Lipe, 1958 *Excavations, Glen Canyon Area*, University of Utah Anthropological Papers no. 44 [Glen Canyon Series no. 11] [Salt Lake City, 1960]). The University named it Loper Ruin for Bert Loper who built a cabin near the site in 1908 at the mouth of Red Canyon (C. G. Crampton, *Standing Up Country: The Canyon Lands of Utah and Arizona* [New York and Salt Lake City, 1964], pp. 164–65).

24. Powell originally called the canyon from the Dirty Devil to the San Juan rivers "Mound Canyon," and the stretch from the San Juan to Lee's Ferry "Monument Canyon." He later combined them into "Glen Canyon."

25. Hillers probably learned this term from Powell. The Ute and Southern Paiute vocabulary lists collected by Powell between 1868 and 1873 contain several entries listing "Unkartoweap" as "Red stone land." (Don D. Fowler and Catherine S. Fowler, eds., "The Anthropology of the Numa: John Wesley Powell's Manuscripts on the Numic Peoples of Western North America, 1868–1880," *Smithsonian Contributions to Anthropology*, vol. 14 [1971]).

26. See diary note 24.

27. Jones ("Journal," p. 98) lists the prospectors as George Riley and John Bonnemort.

28. Hillers may have heard this "tall tale" from Riley or Bonnemort. The Dominguez-Escalante party of 1776 contained only ten persons, Dominguez and Escalante being the only priests. The purpose of the party was exploratory and not to distribute priests among the Indians. The party returned safely to Santa Fé in late 1776. See Herbert E. Bolton, "Pageant in the Wilderness: The Story of the Escalante Expedition to the Interior Basin, 1776, including the Diary and Itinerary of Father Escalante," *Utah Historical Quarterly*, vol. 18 (1950).

29. Hillers's diary for October 10 through November 30, 1871, is the only extant account of Powell's trip to Salt Lake City to bring Mrs. Powell, his infant daughter, and Ellen Thompson to Kanab for the winter.

    Meanwhile, Thompson and the rest of the party continued, arriving at the mouth of the Paria River on October 28, 1871. On November 6 they started overland to Kanab, arriving there on November 12, 1871 (Thompson, "Diary," pp. 56–61).

30. John D. Lee established a ranch on the Skutumpah Terrace in the fall of 1870 shortly before he was excommunicated by the Mormon Church for his alleged part in the Mountain Meadows massacre. With Levi Stewart of Kanab, Lee established a sawmill in Tenney Canyon, five miles from Skutumpah, but sold his interest in the mill in the spring of 1871 (Juanita Brooks, *John Doyle Lee: Zealot, Pioneer, Builder, Scapegoat* [Glendale, 1962], pp. 287–97).

31. Circleville was abandoned in June 1866 during the so-called Black Hawk War of 1865–68 (Peter Gottfredson, ed., *History of Indian Depredations in Utah*... [Salt Lake City, 1919], pp. 145–47, 176, 220; Gustive O. Larson, *Outline History of Utah and the Mormons 1776–1896* [Provo, Utah, 1958], pp. 169–70).

32. The Latter-day Saints Church records show that Archibald W. Buchanan was Bishop at Glenn's Cove in 1871 (Marilyn Seifert, personal communication).

33. Hillers is probably referring here to attempts by the federal government, led by Judge James B. McKean, to suppress polygamy (Ray B. West, Jr., *Kingdom of the Saints: The Story of Brigham Young and the Mormons* [New York, 1957], pp. 316–24).

34. It is not clear for whom this party was working. It may have been attached to the United States Geological Exploration of the Fortieth Parallel led by Clarence King (William H. Goetzmann, *Exploration and Empire: The Explorer and the Scientist in the Winning of the American West* [New York, 1966], pp. 430–66; Richard A. Bartlett, *Great Surveys of the American West* [Norman, Okla., 1962], pp. 123–215).

35. Hillers's chronology is confused here. The section of the diary dated between November 19, 1871, and February 2, 1872, was apparently written on or after February 3, 1872, and Hillers remembered some dates incorrectly. Other diaries of the Powell expedition indicate that Powell and his party arrived at Johnson on November 30, 1871; the camp was moved to Eight Mile Spring on December 3. Bishop, Dellenbaugh, Clem Powell, Hattan, and Riley arrived from House Rock Valley on December 5, 1871. Thompson moved to a camp four miles south of Kanab on December 7 (Thompson, "Diary," pp. 62–64; W. C. Powell, "Journal," p. 372).

36. Powell had hired George Otis ("Bonny") McEntee in Salt Lake City (W. C. Powell, "Journal," p. 372, n. 90).

37. The Powells hired Lavina Nebeker as a nursemaid to help care for their new daughter.

38. John F. Steward had been ill during the latter stages of the river trip; he left to return to the East (John F. Steward, "Journal of John F. Steward," ed. William G. Darrah, *Utah Historical Quarterly*, vols. 16–17 [1948–49]: 250–51).

39. Stewart was the son of Levi Stewart of Kanab.

40. Powell and the men started from Thompson's camp on December 22, 1871, returning on January 2, 1872. (Thompson, "Diary," p. 64).

41. Powell had decided to return to Washington to seek further appropriations for continuing the Survey. He was successful and returned to Kanab in early August, 1872. Thompson, meanwhile, remained in charge of the party.

42. Thompson ("Diary," p. 67) records for January 31, 1872: "Went to Kanab with the Major. He settled up with Beaman." W. C. Powell ("Journal," p. 395) wrote on February 4, 1872, after he had reached Kanab, "We found that the Maj., wife, baby, Vina, and Jack had started to Salt Lake City on the way to Washington and that Beaman was discharged, the Maj. and Prof. being displeased with him." Beaman ("The Cañon of the Colorado, and the Moquis Pueblos," *Appleton's Journal* 11 [1874]: 548) puts things in a different light: "I with regret severed my connection with an undertaking that had my warmest sympathy." The archives of the Office of the Secretary of the Smithsonian Institution contain various receipts relating to the Powell

expedition, among them one reading: "U.S. Topographical and Geological Survey of the Colorado River, J. W. Powell, A. H. Thompson. Jan. 31, 1872. For services rendered as Photographer from Apr. 1, 1871 to Jan. 31, 1872, 10 months, at 80 per mo. $800.00, Rec'd payment in full, [signed] E. O. Beaman."

Beaman went to Salt Lake City, procured a pack outfit, and made a trip to the Hopi mesas (see Beaman, "The Cañon of the Colorado").

43. It is not clear to whom Hillers is referring here; he may have meant William Thompson who was in later years a federal marshal at Beaver (Juanita Brooks, personal communication).

44. Thompson ("Diary," p. 69) calls him "Alfred Zenng"; Jones ("Journal," p. 111) calls him "Alfred Young."

45. W. C. Powell had been Beaman's assistant throughout the river trip. He had much trouble with the complicated wet-plate photographic process. Thompson ("Diary," p. 68) attempted to help him but with little success. Clem was eventually replaced by James Fennemore and later, by Hillers.

46. "George Adair commenced work at $40.00 per month" (Thompson, ibid.).

47. "Captain Dodds and Jones started to go as far on Kaibab as they could" (Thompson, "Diary" p. 69).

48. That is, glass for photographic negatives.

49. W. C. Powell's diary for this period is missing. A letter by him published in the Chicago Tribune July 11, 1872, contains only this note: "On the 3rd of March I started out, with Assistant Jolly Jack to take views."

50. Johnson was a school teacher in Kanab and had become well acquainted with F. M. Bishop (Darrah, Powell of the Colorado, p. 449).

51. Thompson ("Diary," p. 71) records Jones and Johnson as leaving for "Signal Station" on March 12, 1872 and returning the following day.

52. Fennemore was an assistant in the Savage and Ottinger gallery in Salt Lake City. Powell convinced him to join the survey; he remained until mid-August, 1872.

53. The fort at Pipe Springs, constructed under the supervision of Bishop Anson P. Winsor (see W. C. Powell, "Journal," p. 400).

54. Clem valued the rifle, a Winchester, highly. It was given to him by the Major. He found it after a day and a half search (W. C. Powell, "Journal," p. 402).

55. Named for Lyman Trumbull, U.S. Senator from Illinois, 1854–73.

56. The men were searching for access routes through which to bring supplies for the projected river trip into Grand Canyon during the coming summer.

57. Powell used the term "Shinumo" to refer to the Hopi Indians, as well as to the Puebloid archeological sites found in the country. Hillers is here following his usage. The term "Moqui" was also used synonymously (see April 1, 1872, entry).

58. Thompson ("Diary," p. 76) includes W. C. Powell in the party and indicates its destination as Sharp Mountain (Mount Bangs).

59. A farm between Toquerville and Rockville established in 1861 by Chapman and Homer Duncan (Juanita Brooks, personal communication).

60. There are no diary entries for May 3–4, 1872. According to Jones's "Journal" (p. 124), Hillers and Fennemore arrived at Pipe Springs on May 4. Jones and Johnson arrived there the following day.

61. The "party" consisted of Thompson, his wife, W. C. Powell, Dellenbaugh, and George Adair. They had made a reconnaissance into the Virgin Mountains southwest of St. George and had returned to Pipe Springs via Fort Pearce (Thompson, "Diary," pp. 76–77; Jones, "Journal," p. 125).

62. Hattan and Dodds had been to the mouth of the Paria River to check on the boats the party had cached there the previous fall (Thompson, "Diary," pp. 76–77).

63. Thompson was preparing to start on an expedition to locate the mouth of the Dirty Devil River and to retrieve the Cañonita which had been cached there in the previous year. The party consisted of Thompson, Dodds, Hillers, Dellenbaugh, Fennemore, Jones, W. C. Powell, Hattan, Johnson, and Adair.

64. See diary note 30.

65. Emma Bachellor Lee was established at Lonely Dell (later Lee's Ferry) at the mouth of the Paria River; Rachel Woolsey Lee was established at Jacob's Pool in the House Rock Valley (Brooks, John Doyle Lee, pp. 303–16).

66. "Indian Tom," a Kaibab Paiute, had agreed at Skutumpah to guide the party to Potato Valley (Jones, "Journal," p. 128).

67. That is, the head of Henrieville Creek (Thompson, "Diary," p. 80).

68. Albert Bierstadt (1830–1902) who was famous in the nineteenth century for his grandiose landscapes of the Rocky Mountains and the West (John G. Ewers, Artists of the Old West [Garden City, 1965], pp. 174–85).

69. Upper Potato Valley is not the headwaters of the Dirty Devil River, but a part of the drainage of the Escalante River as Thompson ("Diary," p. 81) concluded three days later (see below).

70. The site at White Canyon (see diary note 22).

71. The Loper Ruin (see diary note 23).

72. Fennemore had been sick for some time. In his diary, John D. Lee recorded that Fennemore "now in the employ of Maj. Powell at $100, per Month, became quite feeble through exposure, being of a delicate constitution, which rendered him entirely unfit for the laborious duties…involving [sic] upon him" (quoted in Brooks, John Doyle Lee, p. 312).

73. Hillers apparently is referring here to the series of incidents and depredations during the spring and summer of 1872 in the San Pete Valley, Utah Valley, and around Nephi and Richfield, Utah (Gottfredson, History of Indian Depredations, pp. 294–307). The incidents included Uintah as well as other Ute groups in the area.

74. Harvey C. DeMotte was a professor of mathematics at Illinois Wesleyan University. Powell had invited DeMotte to accompany him to Kanab during the summer of 1872. DeMotte had agreed to help in determining the exact longitude of Kanab for mapping purposes. Between August 6 and 13, 1872, Powell, Thompson, DeMotte, and others made a traverse across the Kaibab Plateau on their way to the mouth of the Paria River. DeMotte reported his experience in a series of articles in the *Illinois Wesleyan Alumni Journal* in 1873; the articles are reprinted in Elmo Scott Watson, ed., (*The Professor Goes West: Illinois Wesleyan Reports of Major John Wesley Powell's Exploration, 1867–1874* [Bloomington, Ill., 1954], pp. 63–107).

75. Kwagunt was a Southern Paiute Indian. As a young child, he and his sister had reputedly been the only survivors of an attack (presumably by Yavapai Indians) on his family's band then camped on the Kaibab Plateau. The children somehow made their way to another band camped near what is now Kanab. Kwagunt Hollow on the Kaibab Plateau is named for him (see Brigham A. Riggs, "The Life Story of Quagunt, a Paiute Indian, told to Brigham A. Riggs, a cattleman of Kanab, by the Indian himself," MS on file, Bancroft Library, University of California, Berkeley).

76. Fennemore returned to Salt Lake City with DeMotte and Francis M. Bishop, arriving there on August 28, 1872, according to the *Deseret News* for August 30, 1872 (quoted in Watson, *Professor Goes West*, p. 105).

77. See Beaman, "The Cañon of the Colorado."

78. The words "red sandstone" are written above "shales" in the notebook.

79. Powell and Jones had gone with Chuarumpeak and another Kaibab Paiute man to upper Kanab Creek and Long Valley. From there, Powell, Jones, and Joseph W. Young, a resident of Long Valley, hiked and waded their way down the East Fork of the Virgin River through Parunuweap Canyon to Shunesburg (Jones, "Journal," pp. 160–61). They were apparently the first white men to make the hike.

80. Thompson ("Diary," p. 100) records these events as occurring on September 29.

81. There are no diary entries for October 3–8, 1872. W. C. Powell, "Journal," pp. 457–58, indicates the time was spent getting ready for the trip to the Hopi mesas and in taking photographs of the Kaibab Paiute Indians camped near Kanab. Clem also noted that "Jack came to a definite settlement about his wages with him [J. W. Powell]."

82. This incident is recorded in James A. Little, *Jacob Hamblin among the Indians* (Salt Lake City, 1966), pp. 73–74.

83. The village of Hano on "First Mesa" (the easternmost of the three Hopi mesas) is occupied by Tewa-speaking Indians who migrated to the Hopi mesas in the seventeenth century from the Rio Grande Valley during the Pueblo uprising against the Spanish (Edward P. Dozier, *Hano, a Tewa Indian Community in Arizona* [New York, 1966], pp. 1–10).

84. The entries for 1872 end at this point. Hillers, W. C. Powell, Jacob Hamblin, and the rest of the party remained on First Mesa for several days, trading and making photographs of the Indians and their villages. The party then moved to Mishongnovi on Second Mesa and later returned to Kanab on November 11, 1872 (W. C. Powell, "Journal," pp. 464–71).

On November 30, 1872, Hillers, John Wesley Powell, Walter Clement Powell, Andrew Hattan, Stephen V. Jones, and Joseph Hamblin started for Salt Lake City. Thompson, his wife, and Dellenbaugh remained in Kanab (Thompson, "Diary," pp. 106–7; W. C. Powell, "Journal," pp. 474–77).

In the fall of 1873, Hillers accompanied Powell and G. W. Ingalls from Kanab to St. George and on to Las Vegas to take pictures of the Paiute Indians (see Introduction).

Throughout 1873–74, Thompson, Hillers, and others worked to map the High Plateau area of central Utah (J. W. Powell, *Report of Explorations in 1873 of the Colorado of the West and Its Tributaries* [Washington, 1874]; J. W. Powell, "Survey Under Professor Powell," *Smithsonian Institution Annual Report for the Year 1874* [Washington, 1875], pp. 40–42).

85. The entries for "Sept. 11" through "Sept. 15" are on a loose sheet, which has no year date. It is apparently 1873, since a manuscript catalog of Hillers's photographs lists several views taken near Gunnison, Utah, in that year ("Catalog of Negatives, River, Land, and Ethnographic, 1871–1876," Bureau of American Ethnology Manuscript Collection, Smithsonian National Anthropology Archives, Washington, D.C.).

86. Hillers was assigned by Powell to go to Indian Territory and make a series of photographs to be used in the Smithsonian Institution and Bureau of Indian Affairs exhibits at the Centennial Exposition in Philadelphia in 1876.

87. Ingalls had previously been an Indian agent in Utah. He, Powell, and Hillers had attended conferences with various Utah Indians in 1873 (see diary note 84).

88. Okmulkee, or Okmulgee, was designated the capital of the Creek nation in 1867. Beginning in 1870 Okmulkee was the meeting place for the General Council of the Indian Territory which represented all the nations and tribes in the territory (Muriel H. Wright, *A Guide to the Indian Tribes of Oklahoma* [Norman, Okla., 1951], p. 138). General Shanks, a Special Commissioner for the Bureau of Indian Affairs, addressed the Council on May 5, 1875 (Journal of the Sixth Annual General Council of the Indian Territory, p. 12 [copy in the Indian Archives Division, Oklahoma Historical Society, Oklahoma City]).

89. It is not clear what "Tocopes" means.

90. Wewoka was established as the capital of the Seminole nation in 1868 (Wright, *Guide to the Indian Tribes*, p. 235).

91. This is probably John F. Brown, known as Governor Brown. He was the son of a Scottish physician, Dr. John Frippo Brown, and a Seminole woman. Dr. Brown had attended the Seminoles during their move from Florida to Indian Territory. The son, John F. Brown, was a partner in the highly successful Wewoka Trading Company (Wright, *Guide to the Indian Tribes*, pp. 234–35). Hillers's references to Brown's "English nobleman" father and his being a "trader" strengthen the inference that this is "Governor" Brown. One source, however, indicates that Dr. Brown was a "South Carolinian of Scottish descent" (Mrs. William S. Key, "Tribute to Alice Brown Davis." A copy of this address is in the Indian Archives Division, Oklahoma Historical Society, Oklahoma City.).

92. Philip McCusker was a white married into the Comanche tribe who was regarded as its official interpreter (A. A. Taylor, "Medicine Lodge Peace Council," *Chronicles of Oklahoma* 2 [1924]: 106–7).

93. John Chupco (d. 1881) was the leader of the "loyal" faction of the Seminole Indians, i.e., the group that sided with the Union during the Civil War (Wright, *Guide to the Indian Tribes*, pp. 234–35).

94. John Jumper (d. 1896) was leader of the faction sympathetic to the Confederacy in the Civil War. He organized the Seminole Battalion of the Confederate Army, achieving the rank of colonel (Wright, *Guide to the Indian Tribes*, pp. 234–35; Carolyn T. Foreman, "John Jumper," *Chronicles of Oklahoma* 29 [1951–52]: 146–47).

95. Jones was a Baptist missionary and had earlier been an Indian Agent to the Cherokee in 1872–73 (*Report of the Commissioner of Indian Affairs for 1873*, p. 294).

96. A photograph of the seminary, not by Hillers, is contained in Wright (*Guide to the Indian Tribes*, p. 71). The seminary for women, a second seminary for men, and the orphan asylum at Fort Gibson were built and operated by the Cherokee nation (S. W. Marston, "Report of the Union Agency, I.T. to the Commissioner of Indian Affairs, for 1875–76," *House of Representatives, 44th Congress, 2nd Session, 1876–77, Executive Documents* I, part 5, p. 465).

97. "…now the [Cherokee] legislature assembles in a spacious brick council house…which cost in erecting the sum of $22,000" (Marston, "Report of the Union Agency").

98. William Potter Ross (b. 1820) was elected principal chief of the Cherokee in 1872. He was the nephew of Chief John Ross who led the Cherokee from 1828 to 1866 (Grace S. Woodward, *The Cherokees*, The Civilization of the American Indian Series no. 65 [Norman, Okla., 1963], p. 318; Wright, *Guide to the Indian Tribes*, pp. 62–72).

99. Pickering was the Indian Agent at the Sac and Fox agency.

100. Whistler was also a trader at the Sac and Fox agency.

101. Possibly this was W. L. Austin, who is listed as a trader from late 1875 to 1877 in the Sac and Fox agency records (Rella Looney, personal communication).

102. Hillers is referring here to the engravings based on his photographs used to illustrate three articles which John Wesley Powell did for *Scribner's Monthly* in 1875–76. The first two articles may have been available by May 1875. See John Wesley Powell, "The Cañons of the Colorado," *Scribner's Monthly*, vol. 9 (1874–75): 293–310, 394–409, 523–37, and, "An Overland Trip to the Grand Cañon," *Scribner's Monthly*, 10 (1875): 659–78.

103. The diary ends abruptly at this point and we have no further indication of the outcome of this *tête-à-tête*.

104. Raymond S. Brandes, *Frank Hamilton Cushing: Pioneer Americanist*, Ph.D. diss., University of Arizona, Tucson (University Microfilms 65–9951).

105. William G. Darrah, *Powell of the Colorado* (Princeton, 1951); Wallace Stegner, *Beyond the Hundredth Meridian: John Wesley Powell and the Second Opening of the West* (Boston, 1954); Donald Worster, *A River Running West: The Life of John Wesley Powell* (New York and Oxford, 2001).

106. For Hillers's photographs exhibited at the 1876 Philadelphia World's Fair, and his work thereafter in the Southeast, Southwest and California until his retirement, see Don D. Fowler, *The Western Photographs of John K. Hillers: Myself in the Water* (Washongton D.C., 1989).

PART 3

Francis Bishop's 1871 River Maps

Francis M. Bishop

Map-making was considered important to Powell on both river expeditions, although the means employed in map preparation necessarily fell short of the optimum in engineering precision.

On the 1869 expedition, Major Powell possessed practically a one hundred per cent monopoly on scientific or technical capability. Of the other men, who were apparently chosen for their ruggedness and ability to survive in the wilderness, probably not one had ever before read a barometer-altimeter or had prepared a highly detailed map. But since Powell could not do everything, he instructed O. G. Howland, a newspaperman from Denver, in the rudiments of map-making. As is well known, O. G. Howland was one of the three men who failed to survive the journey. He was killed by Indians when the three struck out overland from the canyon toward the Mormon settlements. Powell, in his diary, stated that the records of the expedition were kept in duplicate and that one set was sent out with the ill-fated Howland brothers and Bill Dunn.

When Captain Francis M. Bishop took over as map-maker for the 1871 expedition, he obviously had with him the surviving copy of Howland's 1869 map. This fact is apparent from Bishop's frequent diary references to the campsites of the first trip—campsites that he could have pinpointed only by reference to Howland's map. Just what Bishop or Powell did with Howland's map after the 1871–72 trip was over will probably never be known, since the map has not been located and there are no written references to its disposition.

Even the official copy of Bishop's map has been either lost or thrown away.[1] Possibly the map had little usefulness to either the Smithsonian Institution or to Powell, since Professor Almon H. Thompson and his men soon made maps of the Colorado Plateau that were far superior to Bishop's. Fortunately, Bishop may have had some inkling of the insecure nature of government files, for when he was at Kanab he wrote in his diary entry for Monday, April 8, 1872: "Am through with my map; have only to trace out a copy for myself and then I shall pack it off for Washington."[2]

When his son, Dr. W. DeLance Bishop, donated Bishop's private papers, journals, and miscellany to the Utah State Historical Society in 1951, among the items were six map sheets covering parts of the Green and Colorado rivers. Since Bishop sent his official maps to Washington, these six sheets at the Society are undoubtedly his personal copies.

Of the ten participants in Powell's 1871 river expedition, three specialized in topographic mapping. They were Almon H. Thompson, Stephen V. Jones, and Francis M. Bishop. A fourth, Frederick Dellenbaugh, began the voyage as an artist, but later was considered an aide to the topographers. Undoubtedly, Thompson was in charge of the map-making phase, not only because he was second-in-command of the whole expedition, but also because he was a competent scientific leader.

Stephen V. Jones is variously described as topographer and as assistant topographer. Throughout his diary, Jones wrote about preparing maps of the river, but his graphic products seem to have been lost, or possibly thrown away, some time after the voyage.[3]

Thompson himself probably did not prepare any maps while on the voyage, but rather confined his efforts to directing the work of Jones and Bishop. In the years immediately after the trip, however, when Thompson prepared detailed topographic maps of the entire area, he probably referred to the river topography as set down by Jones and Bishop.

Dellenbaugh described the method of taking the topography.

> It was the duty of Prof. [Thompson] and Jones to make a traverse (or meander) of the river as we descended. They were to sight ahead at each bend with prismatic compasses and make estimates of the length of each sight, height of walls, width of stream, etc., and Cap [Bishop] was to put the results on paper.[4]

As a result of this teamwork, Bishop's was probably the main map, whereas Jones's map may have served as a backup.

Although Bishop's maps are the first reasonably accurate maps ever drawn of the Green and Colorado Rivers, and although they are now available to peruse, two drawbacks limit their usefulness. First, only six sheets are extant, whereas a complete set would probably include ten or eleven sheets. Whirlpool, Split Mountain, Desolation, Gray, and Labyrinth Canyons, and the upper part of Glen Canyon are missing. (Grand Canyon is not shown because Bishop left the group at Lee's Ferry.)

The second drawback is the small and cramped writing on the map. Because of this small writing, only sample sections of the maps have been printed full size to show the appearance of the map and to suggest the type of information contained thereon.

One must not conclude that because the maps are sketchy and somewhat inaccurate, Bishop and Thompson lacked skill. A thorough triangulation mapping survey is a painstakingly slow operation and was out of the question for either of the Powell expeditions. What Bishop came up with as features on his map—a snake-like double line indicating the river, a few tributaries, plus some indication of prominent formations and major rapids—is about all that could be expected. That Bishop also entered campsites and noon stops, with dates, can be considered a bonus to historians.

Notes

1. At the editor's request searches for Bishop's "official" maps were made at the National Archives, the Smithsonian Institution, the Library of Congress, and at offices of the U.S. Geological Survey. No maps from the Powell river trips were located.

2. Charles Kelly, ed., "Captain Francis Marion Bishop's Journal, August 15, 1870–June 3, 1872," *The Exploration of the Colorado River in 1869, Utah Historical Quarterly*, vol. 15 (1947), 229.

3. Herbert E. Gregory, ed., "Journal of Stephen Vandiver Jones, April 21, 1871–December 14, 1872," *The Exploration of the Colorado River and the High Plateaus of Utah in 1871–72, Utah Historical Quarterly*, vols. 16–17 (1948–1949).

4. Frederick S. Dellenbaugh, *A Canyon Voyage: The Narrative of the Second Powell Expedition* ... (New Haven, 1961).

All maps are from the year 1871. North is at the top of each page.

Green River, Wyoming, launch site to Black's Fork at center.

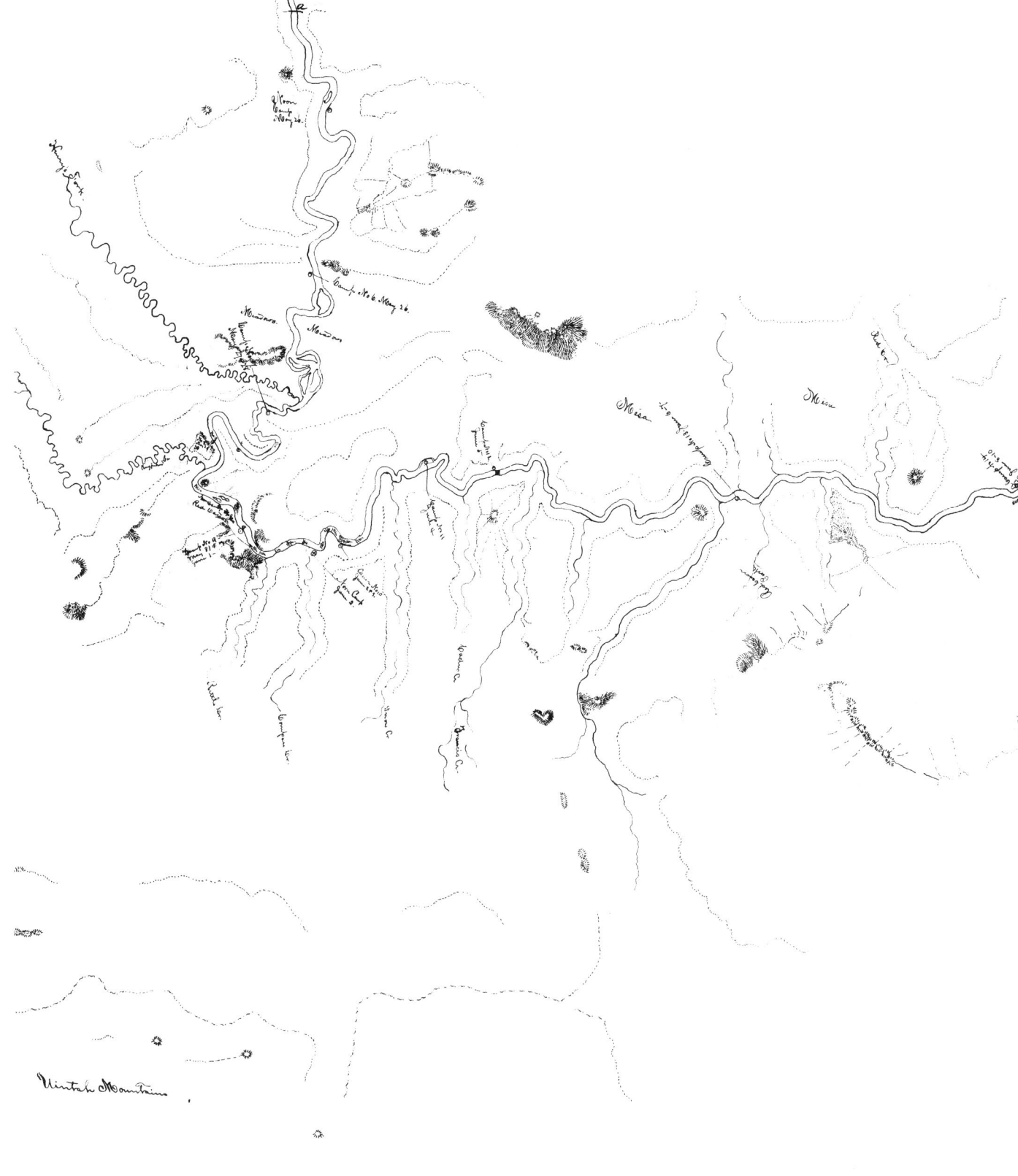

Section through lower end of Red Canyon, Brown's Park, and Lodore Canyon, Green River.

Confluence of the Duchesne, White, and Green Rivers, Uintah Basin.

Confluence of the Green and Colorado (Grand) Rivers (top left), and most of Cataract Canyon.

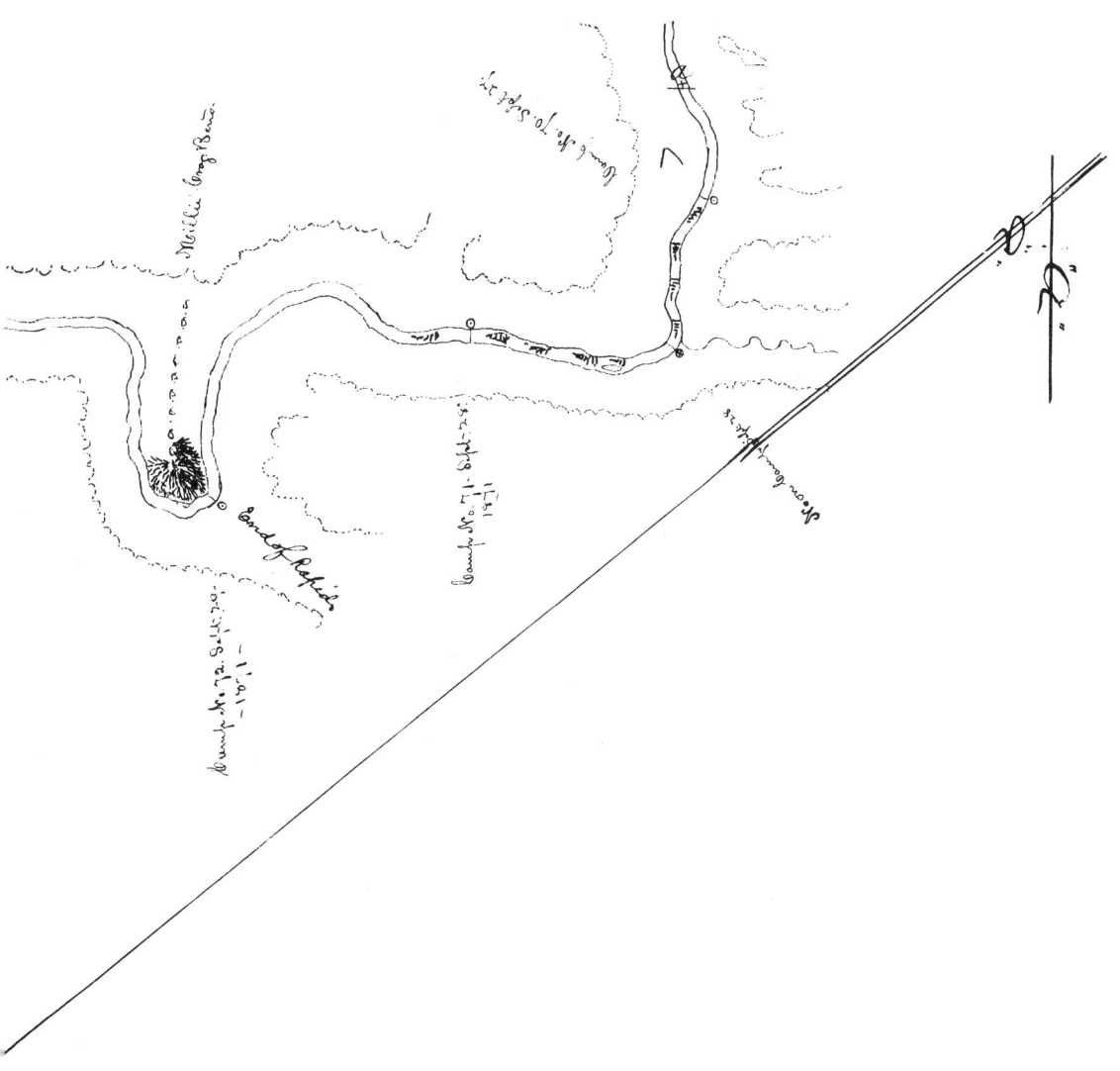

Mille Crag Bend, Narrow Canyon, and part of Glen Canyon, Colorado (Grand) River.

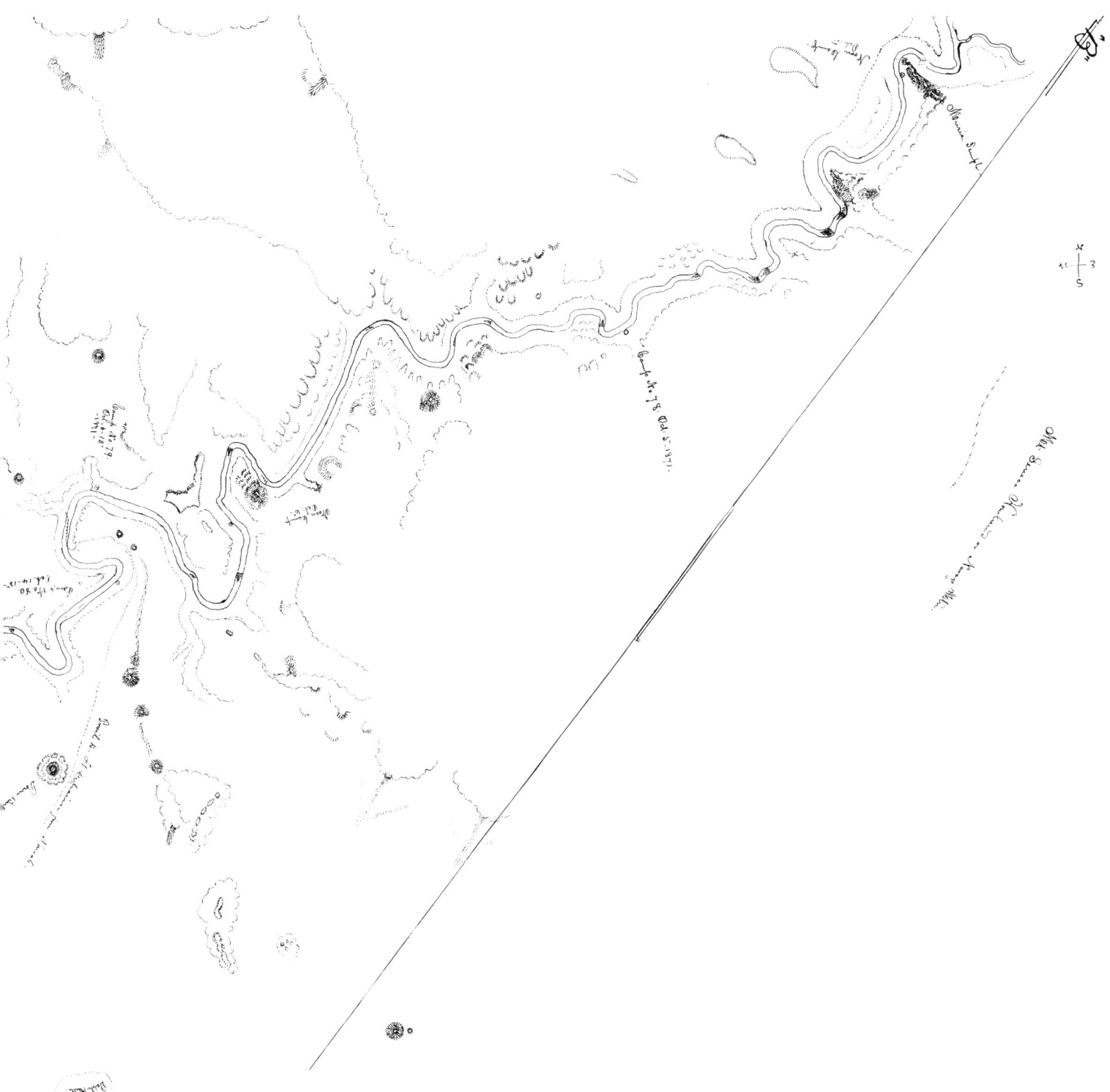

Music Temple (top right), Glen Canyon to mouth of the Paria River (Lee's Ferry) (bottom left).

PART 4

F. S. Dellenbaugh of the Colorado

Some Letters Pertaining to the Powell Voyages
and the History of the Colorado River

Frederick Samuel Dellenbaugh

Introduction by C. Gregory Crampton

John Wesley Powell always thought of his voyages down the Colorado and of his subsequent explorations of the adjacent country as investigations made not in the name of adventure but in the spirit of science. As a consequence his publications are mainly scientific in character—he did not write for large audiences. It was almost with reluctance that he produced a few popular works, the most important of which was his *Canyons of the Colorado* (largely a gathering of earlier pieces) published in 1895 and reissued by the Dover Press in 1961 under the title *The Exploration of the Colorado River and Its Canyons*.

And thus it was with most of those who came under Powell's direction. Unhappily one may look in vain for autobiographical and popular accounts written by Dutton, Gilbert, Thompson, Holmes, or Moran. Alone of all those who rode the Colorado with Powell and who later worked for him in the field is F. S. Dellenbaugh who turned to popular and historical writing.

Frederick Samuel Dellenbaugh (1853–1935) was born in McConnelsville, Ohio, and died of pneumonia in New York City. In a statement written (and preserved in the New York Public Library) when he was eighty-one, Dellenbaugh found a certain continuity in the major events of his life. He learned as a youth how to handle a rowboat in the rapid waters of the Niagara River; he devoured Lieutenant Ives's *Report Upon the Colorado River of the West Explored in 1857 and 1858* (Washington, D.C., 1861) and when the opportunity came at age seventeen to join John Wesley Powell in an exploration of the Colorado River he fell back on his boating experience and his knowledge to make the most of the chance.

"Fred" was the youngest member along on Powell's second voyage of exploration of the canyons of the Colorado River carried out in 1871–72, and he stayed on in 1873 with the Powell Survey of country adjacent to the Grand Canyon region. These experiences were among the most important in his life, and they are of the most interest to those who have studied Powell and his explorations in the West.

Thereafter Dellenbaugh traveled in the West, took up the formal study of art in Europe, and spent much of his life in painting, lecturing, writing, and further travel. Much of this later activity consisted in the development of his experiences with Powell and this is most notable in his writings. With his book *The Romance of the Colorado River* (New York, 1902), he staked out a field as historian of the river. The book, a review of the centuries of history from the coming of the Spaniards to the Stanton Railroad Survey, has not been replaced. In it the author gave a large space to the second Powell voyage which Powell in his own account of the explorations had, oddly enough, not mentioned. Other titles, *The North Americans of Yesterday* (New York, 1900) and *Breaking the Wilderness* (New York, 1905), reflected Dellenbaugh's strong interest in the Indians and the general exploration of the American West.

In 1908 Dellenbaugh published *A Canyon Voyage* (later reissued by Yale and more recently in paperback) which surely established his reputation as historian of the river. The book is an elaboration of his diary of the 1871–73 explorations with Powell. It was the first full and detailed and human account of the second voyage. Now that virtually all of the diaries and contemporary documents of value pertaining to the expedition of 1871–73 have been published (largely by the Utah State Historical Society in its Quarterly in 1947–1949), *A Canyon Voyage* remains as one of the best.

Dellenbaugh's continuing interest and writing about the West, the Colorado River, and Powell (whom he stoutly defended against all critics) elevated him to a position of authority on these subjects, and he was sought out by two generations of historians and other interested persons for information. Up to his death in 1902, John Wesley Powell remained somewhat reluctant to supply detailed information to those who wanted to navigate the canyons but Dellenbaugh complied with requests.

His *A Canyon Voyage* has served more than one party (the Kolbs for example) as its guide to canyons of the Colorado. Like Dellenbaugh, after completing his fantastic railroad survey in 1889–90, Robert B. Stanton turned to an interest in Colorado River history and he, too, came to Dellenbaugh for much help. Only a portion of his work edited by Dwight L. Smith, *Down the Colorado* (which is a summary of the railroad survey), has been published (Norman, Oklahoma, 1965). Along with many others, Dellenbaugh was brought in 1929 to testify in the "River Bed Case" (U.S. *vs.* Utah, a suit before the Supreme Court to determine the navigability of the Colorado River within the State of Utah as a means of determining title to the bed of the river). His testimony was long and detailed and constitutes an important mass of information about the second Powell voyage.

The "River Bed Case" did much to awaken historical interest in the Colorado, and a new generation represented nicely by Russell G. Frazier, Raymond T. Stites, and Charles Kelley came to Dellenbaugh for information about the river. Dr. Russell Frazier's interest in river navigation began when he had a boat on a river near his home in West Virginia. After locating in Utah he planned a boat trip down the Colorado River, which he accomplished in a three-year period beginning in 1933. In preparation for this expedition, he contacted Julius F. Stone, who had navigated the Colorado River in 1909. When Frederick Dellenbaugh came to Utah to testify at the Colorado River trial in 1921, Frazier entertained him and they kept up a correspondence on river matters from that time on. On one of his expeditions Frazier carried a copy of the *Utah Historical Quarterly* containing an article on "The Mysterious 'D. Julien,'" by Charles Kelly, hoping to find some of the inscriptions described. A boat was wrecked on this trip and the copy of the *Quarterly* remained under water until rescued the next year at low water.

Raymond T. Stites was a book collector with a particular interest in any published material on the Colorado River. He collected all of Mr. Dellenbaugh's writings and corresponded with him on river matters until Dellenbaugh's death. He never made a river voyage but was acquainted with most of the river runners who were still living. His collection of river material was very comprehensive and was sold to an eastern book dealer after his death.

Charles Kelly became interested in western history soon after arriving in Utah in 1919. After finding an inscription of Denis Julien, dated 1831, in Uinta Valley, he did research on the French-Canadian trapper and wrote his findings in the *Utah Historical Quarterly*. He also collected much material on the river for his library. When Dr. Julian Steward, of the University of Utah, invited him to go along on an archeological expedition through Glen Canyon he accepted and became much interested in river navigation. Later Dr. Frazier invited him to accompany his various expeditions on western rivers. When Frazier conducted Julius F. Stone on another trip through Glen Canyon in 1938, Kelly was an invited guest. In the meantime he had questioned the accuracy of the U.S. Geological Survey maps on Glen Canyon in regard to the location of the Crossing of the Fathers, and wrote Dellenbaugh to find out the authority for the location at Kane Creek. This started a considerable correspondence with Dellenbaugh on a variety of historical subjects.

The letters that follow have been chosen from many in the files of the Utah State Historical Society written by F. S. Dellenbaugh to these individuals. It is a "one-way" conversation; the Society has only a few of the letters which prompted the replies. In these letters Dellenbaugh straightforwardly supplies the historian with much information. He touches on the Powell voyages and the subsequent Powell Survey; the killing of the Howlands and Dunn by the Shivwits Indians; river navigation; the naming of Grand Canyon and the Painted Desert; the Dominguez-Escalante expedition; Brigham Young and the Mormons; Mountain Meadows Massacre; Jacob Hamblin and John D. Lee; and Robert B. Stanton.

Although he lived most of his life in New York, F. S. Dellenbaugh retained to the end a lively interest in the Colorado River and the West. In addition to the titles already noted, Dellenbaugh wrote two more important books: Fremont and '49 (New York, 1914) and George Armstrong Custer (New York, 1917).

He left behind a wealth of personal papers which reflect his life interests, much of which indeed pertains to the continuity of experience he wrote about in his eighty-first year. His papers are largely located in three places: The Utah State Historical Society; New York Public Library; and the Arizona Pioneers' Historical Society, Tucson.

The Letters

Aug. 29th 1926

Dear Mr. Stites:

The volume "Facts and Figures Pertaining to Utah" "1915" has arrived and I have placed it on my shelves as it contains a vast amount of information on Utah— in which state I may claim a considerable interest since I helped explore a large area of it and have had it "on my mind" for half a century.

The beautiful photographs also came in good order, and the clippings—for all of which please accept my sincere thanks.

Whoever took the photographs knows how. They are exceedingly well done.

I have not read McClintocks "Mormons in Arizona" though I have had it on my notes to do so ever since it appeared. The N. Y. Public Library seems not to have it but perhaps our Explorers Club has as we have a very fine collection of books of that sort. If not, and I can't get either of these to get it, I will buy a copy.

I have the second edition of Jacob Hamblin but not the original. I have used the original—borrowed it from a Salt Lake friend, but had to send it back and I have been unable to persuade him to sell it to me and I can't find another. There are one or two errors in it as to dates. I can't lay my hands on my copy or I could tell you what they are. One, I remember, is with reference to the time when Jacob went with Major Powell across to the Hopi Towns. I think he gives the year as 1871 when it was 1870. If your duplicate is the original edition I decline to rob you, but thank you all the same for your great generosity in either case in offering to send it to me.

I have before me yours of the 23d but I have mislaid for the moment—it is somewhere on my desk—your other letter of recent date in which you mention Prof. Bolton's following out the Escalante Trail. He is good at that sort of thing and he will produce a valuable book. He may get his out before I do mine as I want to go over a portion, or portions, of the trail also—especially in Colorado and up the Strawberry. I know it from Utah Lake around to the Crossing of the Fathers and I could hit it pretty close elsewhere but I like to do "my level best" on it and so will not be in a hurry.

The Crossing of the Wasatch has bothered many but it is perfectly clear to me. I have had a translation made from photostat reproductions of the written copy. I don't know the one Prof. Bolton has used—it may be better. I find only minor descrepancies in comparing with Harris's—not much left out—but as I said Prof. B. may have had access to a copy I do not know about but I will investigate. I think my volume on Escalante will perhaps be more elaborate than Prof. Bolton's—that is as to maps and illustrations—but I don't know his plans. I wish him all good luck possible.

With my regards,
Sincerely Yours
Frederick S. Dellenbaugh.

Oct. 5th 1926

Dear Mr. Stites:

It is most kind of you to send me the Escalante clippings; and just now the Jacob Hamblin reached me in good order. May I venture to say that I think you ought to let me pay for the Hamblin?

I think I told you that I knew Jacob very well, also his brothers Fred and Lyman; and his son Joe. The latter was wonderfully expert as a youngster in throwing the lasso. He could catch any animal by any portion of it he chose. The Hamblins were all sterling, reliable men. They were with our outfit a great deal in the 70's and gave us good work.

I have for several years been trying to have a monument of some sort put up in Kanab to commemorate the fact that that village was headquarters for the Powell survey parties for a number of years from 1871 on. It need not be a fine monument. I wrote to Randall Jones about it the other day and sent him a copy of the inscription the Reclamation Service got up in conjunction with me. At that time Arthur Powell Davis, a nephew of Major Powell's, was director and he was interested. I do not know if the present director is.

I believe at that time the Commercial Club of Salt Lake was also interested.

A good concrete monument four or five feet high on a solid foundation would be sufficient for historical record.

Escalante and his partner Dominguez are coming into their own. The boulder at Spanish Fork is very effective. I hope the name of Dominguez is on it as well as that of Escalante for he was a co-leader evidently.

Thanks for all the things you have so kindly sent me—but I think I ought to pay for the Hamblin.

Sincerely Yours

Frederick S. Dellenbaugh.

You are, I believe, in the D & R. G. Ry offices. I wonder what became of a picture in oils about 30" × 40" of a train rounding Veta Point which I painted in 1877 and gave to Col. Dodge?

July 23d 1929

Dear Mr. Stites:

I have just received the copy of Parry's Monthly Magazine, Vol V—No 6—March 1889, and thank you very much for your kind thoughtfulness.

The article "The Grand Canyon of the Colorado" as you doubtless noted has nothing in it about the Grand Canyon, but probably the earlier part had.

The picture entitled "Running a Rapid, Grand Canyon,["] is from Major Powell's Report but nevertheless is entirely false. The boat just coming over the drop would not have survived a minute in the position given and the one in the foreground is a joke.

The steersman is pulling on a tiller as the boat has a rudder. Our boats had no rudders. No boat on the Colorado had a rudder—that is the ones that ran rapids—for a rudder is useless.

Nobody ever stood up in the bow for they would have been tossed overboard. And the oarsmen never sat so nonchalantly looking aimlessly about. Running rapids on the Colorado is a business that requires all the attention a boatman has.

I don't know why the Major let this crazy picture go through. Perhaps he did not even look at it. It conveys an absolutely wrong impression of the way rapids are run.

I thought you would like to know how bad it is.

I shall have the pleasure of seeing you when I come to Salt Lake next month.

Sincerely Yours

Frederick S. Dellenbaugh.

July 23rd, 1931

My dear Stites:

It was most kind of you to send me that vastly interesting book, SALT DESERT TRAILS, and I thank you heartily for it, BUT, I think you ought to let me pay.

A friend of mine who is a writer was visiting me when it came. He picked it up at random and was deeply interested, reading it steadily till finished.

Recently I found, in an old scrap book lent me, John Moss's own story of his descent through the Grand Canyon in the summer of 1861 from Lee Ferry. He built a raft at Lee Ferry. So he says!

I have heard of this alleged exploit before but never have been able to get the details. He wrote after the publication of Powell's report so he probably gleaned from that and Powell's Scribner articles of 1875. The Moss story came out in the San Francisco Bulletin April 12th, 1877. I wonder if Shepard could find a copy? I have to return the scrap book. I hope to get the owner to give it to the N. Y. Public Library. It contains a lot of newspaper clippings on the two Powell expeditions, letters by Powell, and by his nephew, Clement Powell–NO not nephew–Clement Powell was a cousin or something. I said nephew in one of my books thoughtlessly. The Major had two brothers. Walter was with him on the 1869 trip. He never married. Bramwell was married but his children were very young in 1871—a boy Billy, and Maud who later became famous as a violinist. Bram's wife never knew who her parents were. They arrived with her, a very small child, died of small pox forthwith, and were hastily buried. The only thing Mrs. B. Powell had to trace them by was a small locket and of course never could find out anything. Other persons gathered in the effects of her parents.

I have just had a visit from my grand daughter—aged four and one half—a real little beauty—very sweet tempered, and you can imagine how I miss her voice and antics.

Sincerely yours,

Frederick S. Dellenbaugh.

May 24th, 1933

My dear Stites:

Glad to get the Mountain Meadows "poem." It tells the story interestingly. Thanks.

The only drawback to the Hamblin part is this:

Jacob told me that if he had been at home that day the crime would not have happened. He was absent he said and returned after the event.

There was no necessity for him to tell me that, unless it were true.

I have no doubt that he would have prevented the slaughter.

I met Lee, as you know, saw much of him for a time at the mouth of the Paria. He declared he tried to prevent the massacre but of course that was a lie, as was shown at the trial.

I have a photograph of him sitting on the "rough box" just before he was shot.

I was sorry to read in the Times here that Capt. Bishop died on the 22d.

I sent you a copy of a book published by the Authors Club—an essay by Woodbury on Virgil.

Sincerely Yours

Frederick S. Dellenbaugh.

May 29th, 1933

My dear Mrs Stites:

Let me thank you for your kind thoughtfulness in sending me immediately the obituary on Captain Bishop.

There was a brief mention in the New York Times so I was not unaware of what had happened.

Captain Bishop was so closely associated with my first entrance into the mountains that I could never forget the kindly spirit he always bestowed on me. He was a fine frontiersman and I learned a great deal that was of vast benefit in later years.

In fact that was a wonderful school for me. Between Major Powell, Thompson, Bishop, Hillers, and all the others I had teachers who knew the ropes if any one did. Powell was a marvel in his original way of taking things; and Thompson was about the same. Bishop too never flagged and was always ready with a remedy. So I kept my end up for anyone would have been ashamed not to with such a bunch.

Bishop had to leave us at the end of the first season on account of his health which was a regret all round. I had a hope that I might see him again this year but the fates willed otherwise. I am the last one now of that river party.

With kindest regards,

Sincerely yours

Frederick S. Dellenbaugh

October 11th 1934

My dear Stites (and "Jane"):

It seems a long time since I have had communication with you, so I am "taking my pen in hand" and ignoring the "typer" at my elbow, because writing seems less formal—more intimate—less like a commercial transaction. And I want to send you a letter I wrote one of our local papers on the subject of naming the Grand Canyon. There is still a mistaken idea in some quarters that somebody other than Major Powell, named it.

Positively Major Powell named it and no one else and I put the name on the map I drew as stated in the letter herewith. If I don't know then nobody living does. You will recall that I had to correct the mistaken notion that "the Early Spaniards" named the Painted Desert. It was Lieut. Ives, in 1858, who named that....

With regards

Faithfully Yours

Frederick S. Dellenbaugh

Jan. 23ᵈ 1933

My dear Dr. Frazier:

Your most delightfull and complimentary letter has just arrived and I hasten to assure you that I greatly appreciate it and will preserve it in my "archives."

The voyage through those upper canyons of the Green is a beautiful one—not so tremendous as Cataract and the Marble-Grand, but quite as beautiful and interesting.

Your proposed voyage this year as far as Lee Ferry you will find fascinating. When I was in Los Angeles in 1929 testifying in the U.S. vs–Utah river-bed case, I heard much about "sand-waves" in Glen Canyon. Although I went through Glen twice I saw no sandwaves but they are there at times and you must be prepared for them. The whole bottom seems to rise up suddenly and I am told the proper method is to take them <u>sidewise</u>, strange to say. They are likely to turn a boat end over if taken bow on.

If you should care to get points on these I am sure Air. Dave Rust, Provo, Utah, would give you some. He has gone through Glen many times and is a "corking" fine fellow.

I hope you kept a detailed record of your trip—and will of the forthcoming ones, and if you would be so good, I would be pleased to have copies to file with the other Colorado River manuscript material in the New York Public Library. What you do will be history also in a few years. The N.Y.P.L. has the original Spanish MS. of Castaneda (1596) which first mentions the great chasm of the Colorado, and also has many diaries of men who have descended.

Of course I would be more than glad to see your movies but it seems almost too much for you to send them. I would return speedily. I know Bray of Bray Pictures Corporation who would run them through for me.

Please keep me posted on what you do. I congratulate you on having the nerve and the vigor to make these trips.

With regards,

Faithfully Yours

Frederick S. Dellenbaugh.

Feb. 21st, 1933

Dear Dr. Frazier:—

Your letter of Feb. 11th is received, also the copy of A CANYON VOYAGE.

It will be a pleasure to autograph the book. I will return it in due course. I have three other volumes on hand to autograph for a friend–not all one kind, however. This friend is a young man named Bonney who is so enthusiastic about the West and Southwest that he spends all his spare money on books about those regions. If he goes to Salt Lake I want to give him a letter to you, so that he will visit Bingham and see what a marvellous place it is, as well as having the pleasure of meeting you.

You are very kind to say such nice things about my letters. I wish they were more interesting—but perhaps time with its magic power will endow them.

To go through Cataract Canyon you positively must have your boats made with watertight cabins.

Perhaps these decks and bulkheads could be made adjustable and removed after you come out of Cataract. Even with them on you would have about the same storage room; by way of the hatches. And if the decks are strong, you can stand on them. One advantage of compartments is that in the event of a capsize you do not lose your supplies.

Don't minimize the power of the rapids in Cataract. Take every precaution.

I shall be greatly interested to see how a motor will work in the rapids which run at the rate of 20 to 25 miles an hour. I think the motor will have to be turned off and the boat manipulated without.

Yes, there is one place I would be glad to have a description of from your point of view and a photograph if possible. This is the place described on p. 130 A Canyon Voyage. You may hit a stage of water that will make it far easier than found it in 1871, by our outfit.

Then I would be grateful if you could stop at <u>Music Temple</u>, about two miles below the mouth of the San Juan, on the left, and tell me if the names carved there are still visible. Dr. Oastler was there several years ago and then found them almost illegible.

I will later send you a description of the cliffs opposite the landing place, or a photograph which Oastler took.

And one more place—the <u>Crossing of the Fathers</u>; El Vado de los Padres— where Escalante crossed in 1776 on his way back to Santa Fe.

This is about 35 miles before you reach Lee Ferry. If you will take a look at the U.S.G.S. detailed river maps which they must have at the University in Salt Lake, you could locate the place; The Old Ute Ford, and Crossing of the Fathers, exactly.

The crossing trail entered the river from the right through a very small canyon on the map called <u>K</u>ane Creek. The name should be <u>C</u>ane Creek as it was named, I believe, from the canes which grew there in early days—and when we were there in 1871 and 1872. The error doubtless arose from the place being in Kane County.

In the autumn of 1871 there were little piles of stones marking the course of the trail down the middle sand bar.

And still another place, if you should have the time, This is about <u>three miles UP the Paria</u> on the right hand; there is a long steep slope of sand with a precipice at the top.

It was here that Escalante in 1776 climbed out on his way to the Crossing. The trail at the top works its way through a sort of crevice as I remember it. I would be glad to have a photograph and description.

The reason I mentioned a full sized movie camera was that Mr. Bray remarked several times that it was a pity your film was not standard size.

With kindest regards,

Sincerely Yours

Frederick S. Dellenbaugh

May 29th. 1933

Dear Dr. Frazier:

Thanks very much for your kind thoughtfulness in sending me the obituary of my old friend and companion of the canyon days, Captain Bishop.

His crossing the Range was not a surprise to me for besides being on the verge of 90 he had not been quite himself for several years.

He was a fine character—always dependable in every circumstance. I was very fond of him.

When he became a Mormon I was not surprised either for he had a highly religious and emotional nature and was sure to affiliate with the religion nearest him.

The statement that he made the first map of the Colorado river is not quite accurate because his topographic work was in our first year on <u>Green river</u>.

The <u>first map</u> we made was a year later—after the Captain had resigned and it did not take in the Green river work of the Captain, not going above the San Juan.

The map we made in a tent in Dec. 1872 and Jan. 1873 in Kanab was based on our triangulation from a nine mile measured base line at Kanab and the Captain was not in on this at all. His work was incorporated in the upper map later.

Prof. Thompson laid down the triangulation points and I drew in the topography assisted by John Renshawe a new comer then. This map was soldered in a tin tube which I carried on my back to Salt Lake in Feb. 1873 and sent to Washington. It was there divided into two parts which are reproduced in my CANYON VOYAGE. Bishop's name is on as a topographer because he was chief topographer the year before and had done the river down to the mouth of the Green but this part was not used in the map we made in Kanab which was absolutely the first map of the Colorado river above the Grand Wash.

I mention this because in future years someone might get mixed up on it.

How about your proposed trip down from Green River Utah this year. Look out for Cataract Canyon. Be careful in there and you'll come through all right.

And keep me posted for I am greatly interested.

With best wishes,

Faithfully yours,

Frederick S. Dellenbaugh

Don't neglect life jackets for Cataract and have your boats covered—or decked over.

July 12th, 1933

My dear Dr. Frazier:

Your letter of July 5th, is just here, and I am deeply interested in the proposed trip down the Colorado which you tell me is to begin at Ouray on July 26th.

I will be with you in spirit all the time. You were lucky to leave behind the ones you might have had to "wet nurse" for every man must be able to look after himself.

You will enjoy the descent I am sure. Your equipment seems to be all right as to boats. They should float when full of water with the crew and cargo on board and from, your description I should say your boats will.

Desolation Canyon in its 97 miles has many rapids but with due care you can run or portage without much difficulty. Your five men being husky will be able to easily pull the boats out and run them on drift-wood skids over the rocks where necessary.

Be careful in "lining" not to let a boat get very far out and have it caught the way Eddy's was in a "hole." It stood almost straight up in the air and they had to leave it. I think this was in Cataract. Keep inshore when lining, as much as possible.

When you turn from the mouth of the Green into the Colorado and Cataract Canyon, be on the watch for the first big one, three or four miles down. That's where the Brown party lost their zinc boxes holding almost all their grub.

But that was bad management. The middle part of Cataract is the worst, and there take it easy. Don't hurry. Land often. Note the drift of the main current by throwing sticks in above.

If the water should happen to be at the same stage we had when you get to the rapids just below Dark Canyon, a very narrow deep side canyon coming in on the left opposite a 3000 foot cliff on the right with a long rapid at its base, look sharp, as you run through a narrow place, and then land on the head of the island, keeping in the line of dividing waters. Then drag your boats about half way down on the right hand side and pull out into that channel pulling rapidly across it to avoid the dash against the left hand cliff at the point where the two channels unite.

But probably the water will be higher when you get there in August and you can shoot down the middle easily. I only mention it to put you on your guard for at our stage of water it was a difficult place. That was September. In August you will wonder why I even mention it.

I think it would be a fine thing and absolutely appropriate to put up a monument at Separation Rapid to honor the Powell party there, and especially the Howland boys and Dunn. Powell often spoke of them affectionately.

I have opposed putting the names of the deserters on the Powell Memorial at Grand Canyon because it seemed inappropriate to me, but in any other place it would be all right. The whole outfit were on the ragged edge at Separation from the strenuous work, long exposure and lack of food, and when the granite ran up again it was too much for them. But most nerved up and stuck to the Major who said he would go on if only one stood by him. Otherwise with his one arm he could not have managed a boat.

Powell named a cliff in Lodore after Dunn and was bent on naming Navajo Mountain after Seneca Howland. Thompson and I did not think that wise or appropriate so it was not done.

The Howland brothers and Bill Dunn were not ordinarily quitters. They were fine fellows. The conditions were peculiar. One of the Howlands did not want to leave but stuck to his brother.

As to the Julien inscriptions. Charles Kelly of Salt Lake (whom you ought to know (473 First Ave.) has just sent me the UTAH HISTORICAL QUARTERLY for July 1933 (Room 131 State Capitol) containing an article by him on D. Julien which embodies all I know about Julien and much more that Kelly has dug up. You can easily get that and if it does not give you what you want to know write me and I will reply at once with such information as I have. My Julien file is in New York.

We checked the miles travelled by estimating—two men always sighted down a stretch of river with prismatic compasses & put down each his estimate. These were compared and then checked from time to time by sextant observations. Col. Birdseye said our work was remarkably true all things considered. There are no inscriptions made by the Powell parties except the ones in Music Temple. They are almost gone. Inscriptions by travellers and explorers are valuable and are not in the class of those which are stuck around through vanity. So I should say it would be advisable for you to chisel in your names and date in say, Cataract, in a place accessible and protected from rain and sand.

Kelly has sent me a photograph of one made by Antoine Robidoux in 1837 which is highly interesting. This is in Westwater Canyon, Book Cliffs.

Yes, the rapid with the island was not far above Mille Crag Bend.

The name of one of your party interests me—Mr. Fahrni. My grandfather and grandmother were Swiss and my grandmother's maiden name was Farney as we spelled it. It may be the same name. My grandfather came from Switzerland and settled in N. E. Ohio in 1825. His home before that was a few miles east of Bienne.

There was, a few years ago, a Grand Marshall of France a General Farny. That family I believe came originally from Italy.

Then farewell for the present. I am confident you will make Cataract all right because you are willing to be cautious.

GOOD LUCK AND MY HEARTFELT BEST WISHES FOR YOUR COMPLETE SUCCESS.

Faithfully yours,

Frederick S. Dellenbaugh.

August 23rd, 1933

Dear Dr. Frazier:

It was most kind of you to write me so fully about your canyon voyage of this year. I read your letter with an interest that only one who had been over the route could understand.

I congratulate you on your efficient management. As Stefansson maintains, "Adventure is an indication of poor preparation and management," and I agree to that. On that basis you see you take a medal!

I shall file your letter with my valued Colorado material along with what you wrote me last year on the first leg of your expedition. Your successful journey is a proof of what I have always contended; that success may always be the result on the Colorado when there is proper preparation and due reverence for the "Dragon."

In other words no funny business—the Dragon does not like to be treated humorously.

It was a pleasure to find you agree with Major Powell and his crew in thinking Dellenbaugh Butte an artistic work of the gods of erosion. I believe it is now called "The Anvil" locally. The local people do not know about the first names applied as a rule. For a long time Mount Dellenbaugh on the Shewits Plateau was locally known as "The Butte," although for more than fifty years it was on all good maps as Mt. D.

I met Tom Wimmer in Los Angeles at the time of the trial of the "River Bed Case." There were a lot of old "River Rats" on hand then and it was pleasant to meet them.

Evidently the water in Cataract was just right for you which was fortunate.

Sometimes it is very bad and running through in some places is not advisable. Of course we had to be doubly cautious for we were on very short rations, and no one knew how to get to a settlement at that time, overland.

Thanks for the fern you enclosed. It arrived in fine shape and I prize it and shall always take good care of it. It is the kind called Maiden Hair.

Your rapid No. 60, as I wrote you, I have always specially remembered because our approach to it was dramatic in the late afternoon light; and we came near hitting the projecting rock at the foot.

Evidently your boats take the waves well. Ours being very sharp in the bow— and we went bow first—sometimes cut through a big wave instead of climbing over it.

The Julien inscription in lower Cataract I judge is difficult to find as it is above the water on a cliff with no landing place near. Stanton said it must have been made from a boat.

Charles Kelley [*sic*] of Salt Lake City has carried on some investigations as to the identity of Julien and has succeeded in digging up some valuable data. You ought to know him. He wrote SALT DESERT TRAILS, an excellent book. He lives at 473 First Avenue, Salt Lake—Charles Kelly.

It is interesting to get your impressions of Music Temple. On my second visit in 1872 I had some of the same feeling concerning that first memorable expedition. I put my name and that of Jack Hillers on the face of rock; I don't remember doing any other. I wanted to carve those of Johnson and Fennemore but they opposed it. The fact that the men of the first party, who carved their names there and were killed later, made the similar record rather shaky for a superstitious person. These two were also somewhat disturbed I fancy, later when I announced that I had just been a witness to Major Powell's will which he wanted to send out to his wife by the teamsters. But they were sick, anyhow, so could not have gone into the Grand Canyon with us. We had to leave one boat, the *Nellie Powell*, behind and Powell gave it to John D. Lee who had been helpful to us. With this he started the ferry known by his name.

I see in re-reading your description of the names carved in Music Temple, that I carved more than I remembered. Clem Powell, Bishop, and Steward. But I don't remember if I carved the names on the first or second time I was there, probably on the second, for on the first we were about out of grub and did not linger anywhere. Even on the second visit we were very short for it had taken us longer than we had planned to get to the mouth of the Dirty Devil. Thompson took only enough rations for himself to get back to Potato Valley where he was to be met with a pack train from Kanab.

Your suggestion that the Music Temple ought to be preserved as an historical monument is excellent. Then all the names could be put on a bronze tablet sheltered from the sand storms. It would also be a monument to Dunn and the two Howlands whose names were omitted (with my approval) from the Powell Memorial at Grand Canyon. As they were deserters I did not think they ought to be honored by a place on that particular monument but I am desirous of seeing them recorded at Separation Rapid. I would also like to see their names on a tablet in Music Temple. There ought also to be a tablet at Green River, Wyoming, whence both parties started. Such records are valuable for posterity.

Will you kindly give me a detailed description of the ledge below Cane Creek on which you camped—its width, length and so on. Perhaps I told you that a doubt arises now as to Escalante's crossing at the Ute Ford (Crossing of the Fathers)

The way out on the west side was INTO Cane canyon and out that way. The entrance also to go east.

Yours was an ideal trip, and again I congratulate you.

Faithfully yours,

Frederick S. Dellenbaugh.

June 17th, 1934

My Dear Dr. Frazier:

It has been my intention to write you for some time but I seem to be a terrible procrastinator this year. However I do get around to it from time to time and I certainly want to find out now just when you are to start on your Grand Canyon venture.

I suppose not till September as you said, I believe, that you meant to take it on low water. The only difficulty with September is that the water is apt to be pretty cold—it was very cold with us in Cataract Canyon in September, but that perhaps was unusual.

I note that you have four boats and that is well, for with more boats and more men you have an easier time. The great drawback we had in the Grand Canyon was our small party and our big, waterlogged, boats. As they had dried out for eight or nine months we had to soak them a long time to bring the planks together and that made them lack bouyancy in the water and made them terribly heavy on land in portages.

By the way, I notice it is a practice now to attach lines to both ends of a boat in lining down. This seems to me to be bad practice for it has a tendency unless great care is used in handling the stern line to throw the bow out into the current, a thing absolutely to be avoided at all hazards. Therefore I say only one line should be used and this a bow line. The boat must either drift or be guided by men in the water. When drifting, if in the current where the bow might be caught on the inside by the current a man should be on board with an oar to prevent the boat from taking the diagonal of forces and shooting out into the main river. He plys his oar like a paddle on the outside of the bow.

But you now have had the experience of the upper canyons, and know as much about it as I do. Only, I was very familiar with row boats in swift water as I was always on the Niagara as a youngster, crossing from the Canadian side to Black Rock through a full volume nine mile current.

I am pleased to know that Clyde Eddy is to be with you as he has had a lot of experience in the rapids. Emery Kolb is perhaps the most experienced of all the "River Rats" and he has good judgment.

I was proud to have a boat named for me in your last year's trip, so that makes two on the river so far as Clyde Eddy named one of his after me.

I have enjoyed your films greatly and am sending them back now with many thanks as soon as I get them properly packed.

When you wrote that you were coming East I hoped that you might get as far as New York but I see your eastern limit is Columbus. I expect to go up to my shack in the mountains of Ulster County in a short time and I wish you could extend your journey that far and spend a few days with me on top of my mountain. It may be that even now you are somewhere in the East as you were to leave your home in Utah May 12th. I hope so and that I shall have the pleasure of hearing from you.

Glad you met Fennemore. He is a nice chap and a fine photographer. We were sorry he and Johnson could not go into the Grand Canyon with us. If they had been, we would have had a much easier time. We just needed two more men and they were not to be had.

I am very uncertain as to my plans at present. There is a prospect that I may have to go to Utah, but I hope it will be in September or October—after you finish the Grand Canyon, for I should enjoy seeing your pictures as well as your party and you.

Fennemore, as he probably mentioned, was in Music Temple with Hillers, Johnson and me in 1872, when we were bringing the *Canonita* down from the Dirty Devil. The food question was always so troublesome in those days in that region where there was no game or even where there was game for we did not have time to hunt and were always on the ragged edge for something to eat.

I suppose Eddy will let me know all about the broadcasting for I should be heartbroken to miss it.

I have not seen him for some time but I'll try to have a talk with him before I go to my mountain fastness.

With best regards,

Faithfully yours,

Frederick S. Dellenbaugh.

I will send the Walpi photographs in a day or two.

August 23d 1934

Dear Dr. Frazier:

Congratulations again on your wonderful descent through the Grand Canyon.

Low water is certainly worse than high if one does not want to work like a slave. On very high water in one stretch of about 8 miles we did not stop at all for every rock was buried under the giant billows which were smooth on the surface. You will recall my mentioning that circumstance in my Canyon Voyage.

Well, I'm glad you have had the experience of running the whole line of canyons. There is much that is interesting in the upper ones and Cataract is not an easy job.

The Sockdologer at low water is no doubt more dangerous to run than at high, but, of course, a portage or line-down can be made, if desired, at low water. Stanton got part way down and smashed the side of the "Marie" and halted right there to rebuild her by cutting out the middle part. Then two or three days later she was smashed to "smithereens"

Separation is probably the most difficult rapid on the river on account of the twist in the channel, as I understand it. The three deserters balked at it—or rather two did. The younger Howland simply followed his brother's lead, and according to Jack Sumner, the rapids had become a "holy terror" to the elder Howland. I am glad you put up the placque where it will remain for many years as a memorial to three fine men, two of whom simply lost their nerve. And the nerve of the others wasn't any too vigorous.

Sorry Mr. Farnhi had the bad luck to break two fingers but they will mend and be more or less his "souvenir" of the voyage—a voyage you will all remember just as I do our work on the river so long ago.

I like your decision that "one does not feel like a conqueror when one finishes its canyons, but rather a favored child of the Gods." If my contemplated final edition of <u>A Canyon Voyage</u> comes to being, I think I will quote that somewhere in it.

I shall probably see Eddy when I go back to the city and get some of his impressions. I wonder why he did not tell me he was going with you.

The affair which I thought might bring me to Utah this year, "hangs fire," like most things today, but I hope to get to Utah nevertheless one of these days even if I am almost "one more" than an octogenarian.

I have just been over at Litchfield, Conn. to visit my son's family. He has a summer place there. I enjoyed being with my three grandchildren and my fine daughter-in-law even though my son could not leave his Boston (Waltham) business to come there just now. They live in Boston, winters. My elder grandson is as tall as I, and he is only 13. He rides well, swims well and sails a boat well. So does the second one, and my granddaughter 7 rides as well as any of them.

Kindest regards,

Faithfully Yours,

Frederick S. Dellenbaugh.

Dear Mr. Kelly;

Your letter of Feb. 4th was very welcome and I have had it for reply at my elbow ever since but my procrastination now a days is shameful.

I don't know that the Julien Inscription will appear in the Stanton book but I discovered that there were two photographs, of the one I sent you, in the Stanton collection. I will try to secure one for you from the Stanton heirs.

As to the Connelley—Jed. Smith matter. I had some correspondence over Smith with Connelley and I sent him what I had on the subject, as well as a lot of letters from a nephew of Smith with whom I corresponded. Connelley died and these were returned to me O.K.

Perhaps I can take the subject up with the Connelley heirs. I will try later on. I have definitely routed Jed. Smith, as perhaps I have mentioned before, not down the Virgin as has generally been the case heretofore, but down Meadow Valley Wash and the Muddy to the Virgin. I am certain this is correct. Furthermore C. Hart Merriam arrived independently at the same conclusion. The Muddy and streams of Meadow Valley Wash were Adams River.

Yes, perhaps I have a "soft spot" in my heart for "Old Jacob" as I only think of him as I knew him. He was all right so far as anything occurred while I was in that region. So it is not exactly tolerance on my part in having a somewhat tender regard for him but experience with him and his family over a considerable period.

But if you feel confident that you have other views that ought to see the light, go to it and send them forth. If it drives you to this part of the country it will be our gain. I did not know Porter Rockwell. Probably I saw him, perhaps met him, but at that time his alleged exploits were not known to me.

Remy was a bit unfair in some respects. For instance he berates the people of

Parowan for their treatment of him. This must have been partly his fault for those same people gave the distressed Fremont party a fine welcome and shelter and food. Also Carvalho and others relate pleasant relations.

Yes Lee was ploughing when I went across the Paria that morning to interview him. There was plenty of level land for farming.

We landed from our boat after coming down Glen Canyon one morning (this was the second time I came down—the first was the autumn before when Lee had not arrived. We had cached one boat at the mouth of the Dirty Devil. This we now brought down.)

I went across the Paria alone with my Winchester on my shoulder. Why I had the gun I don't know—not for Lee of course. But the gun was sighted by his wife Rachel and she went into the cabin immediately while I continued on to the left leaving the cabin about three hundred yards on the right—perhaps two hundred. Lee stopped the horses and resting his hands on the plough handles turned his head to look at the new comer. I see on referring to A CANYON VOYAGE I say that Hillers was with me and that must be so but my present recollection is that I was alone. At any rate as soon as Lee understood who I was he was very pleasant and always was while we were there. Yet he sometimes thought we might be trying to capture him.

He had started a dam on the Paria to take out a ditch for his farm and as we had nothing else to do while waiting for orders from Powell we turned in and helped him finish the dam. Several times we had dinner with him and his wife Emma. The latter was a quite young and buxom woman very cordial and a good cook. She had a baby about a year and a half old playing around on the ground in front of the cabin.

Lee may have had a small garden patch nearer the house which he worked with a spade and hoe but there is no question about his ploughing on the farm patch further away. On the U.S.G.S. map the comparatively level land is nearly a quarter of a mile wide by three quarters long. This was not all absolutely level but a considerable part of it was tillable as I recollect the place.

I suppose Hillers <u>was</u> with me that morning when I met Lee. One's memory is not always accurate after 60 years. I remember the incident perfectly all but Hillers' being with me.

Lee was a pleasant enough man and I feel sure that, ordinarily he would have had no murderous intentions. This does not excuse his deed of course. Lee, at one time, hid himself in Havasu (Cataract) canyon with the Havasupai. I heard that Klingensmith and I think one other of the Mt. Meadows ringleaders—perhaps Isaac Haight, hid for a time among the Shewits, and Jack Sumner of Powell's first party, asserted that they instigated the murder of the three Powell deserters. I feel sure there is nothing in this but imagination as it is pretty well established that the deed was the work of only three or four of the Indians—"some no sense" Shewits as Thompson's Shewits guide put it in 1872.

The usual amount of snow in the mountains will put the Colorado on the rampage in June, so look out how you venture!

Sincerely yours

Frederick S. Dellenbaugh

Fanaticism makes demons of us all!

Nov. 6th 1932

Dear Mr. Kelly:

My correspondence has been shamefully neglected this last six months. I should have acknowledged your most interesting letter of Aug 11, with the enclosure of the Crossing of the Fathers photographs, long ago. I notice that if I think of writing on the machine I am apt to put off writing letters. Somehow the typing does not seem intimate. The pen suits me better for letters and I try to make my "pen tracks" legible.

The photographs of the Crossing interested me greatly. I have long wanted such photographs and I have blamed myself many times for not making sketches when I was there in 1871 and also getting our photographer, Beaman, to make some photographs for me looking down the river. At that time the stage of water was such that the line of little stone monuments could be seen above the surface marking the exact route down the bar for a considerable distance. When one of Lieut. Wheeler's parties was there the same year, I think it was, the stones were also visible and someone made a drawing reproduced in the Wheeler report showing them, but the sketch is poor.

I have pasted together your three photographs and have studied the view a good deal, I did not get up on top so I cannot visualize the scene as I saw it—besides 61 years is a long look back.

May I trouble you to tell me just where you stood with reference to the mouth of the little canyon through which the trail entered the water? It was full of canes when we were there and I believe the name on the maps should be "Cane" not Kane Canyon, Kane county of course is named after the "Gentile" Kane who was a good friend of the Mormons, but this little canyon hardly would have been given his name.

At the time we were there getting in and out with horses was difficult. As it has been said—you say so also in your letter—access in now easy, someone must have done some work on the rocks back aways in the pass leading down.

Rereading your letter I note that you could find no trace of blasting to make the way impassable but I am sure the Mormons told me long ago that they had put in blasts somewhere to prevent the Navajos from coming across.

Major Powell went out in 1871 from the Crossing and his party had trouble getting up not far back on the trail. I remember that point very well. Of course as time went on easier ways were found. I will investigate farther. Possibly in the Church records of Kanab stake we might find a description of the attempt to prevent Navajo raids by way of the Crossing.

In the matter of the name of Tickaboo Creek. Tickaboo is Ute. While Hoskaninny was a Navajo perhaps he belonged to Patrish's renegades. Patrish, I believe was a Ute, so the Ute and Pai Ute words were familiar in that region as well as the Navajo. As a rule there were few Navajo terms north of the Colorado the Utes and Pai Utes dominating that region. The Crossing of the Fathers was also known as the "Ute" Crossing. I never heard it called Navajo Crossing. The Navajos seem not to have gone north of the river much till the whites began to settle there and they could steal cattle and especially horses. In 1871 they were crazy to get horses for they had lost almost all they had in the "Navajo War," I traded an old plug we had no further use for to a Navajo for several fine blankets, Ten years later I doubt if he would have accepted the beast as a gift—but he was on foot as many were in those years.

Music Temple is on the left bank not very far below the mouth of the San Juan, Dr. Oastler several years ago coming down with Dave Rust (Oastler—a member of the Explorers Club) located it and made photographs. The names were almost obliterated by the forces of erosion. That rock is soft and the wind hurls the sand with great force.

I was not able, when I was in Kanab in 1929, to find the stone monument that we built as a base for the large transit set up to determine the meridian but it may still be there nevertheless, I did not have much time. We located the site of the old fort built in 1870. Part of the site was washed away in that flood that retrograded from the "Gap" to far up the canyon. I believe they put a marker there afterward—at the site of the fort.

When I was in Kanab in 1903 I don't believe I even looked for the transit base. I was not then interested in our old landmarks. No—I don't remember anything in Stanton's diary about the old dredge he put in. I will look it up. His Colo. Riv. diaries are now the property of the New York Public Library and can be consulted any time. You have to ring a bell to enter the MS room. There is a strong wire grating door which enables the attendant to see who is there before he opens as there are many valuable manuscripts within. I would like a photograph of the dredge wreck to put with the Stanton material in the library. I had all of Stanton's data here for several years. Julius Stone has sponsored a book just out <u>Colorado River Controversies</u> in which, under James Chalfont's editorship Stanton's criticisms of Powell are given as well as Sumner's & Hawkins's narratives of the 1869 trip. Much of Hawkins's stuff is trivial and even ridiculous and I think Stanton was hypercritical. He makes too much of the life preserver incident. Powell having only one arm had a life preserver in the 1869 trip which he does not mention in his report and Stanton censures him in unmeasured terms for keeping it a secret, as he claims, and not imploring Brown to take life preservers. Brown had Powell's report to show him what the river was and he could have consulted me and other men of the 1871–72 crowd but he was too confident that the river had no terrors for him. Brown's expedition was so poorly equipped and so poorly directed that it was in for trouble life preservers or not. It was the worst managed expedition that has ever attempted the descent. Disaster was its fate from the start. The Colorado always punishes contempt.

Powell, according to the Stanton-Sumner-Hawkins accusations changed his character for the 1869 trip entirely, before and after being a different man. Powell had his faults but he was not what they try to make out. Well, this is about enough for this time. I hope you can read it. I write rapidly and therefore the letters are not always clear.

Thank you for your interesting and valuable letter and for the photographs.
Faithfully Yours
Frederick S. Dellenbaugh

Nov. 15th 1932

My dear Mr. Kelly:

Ledyard has sent me several of his latest, very interesting "<u>Ax-I-Dent-Ax</u>" magazines and in one I am pleased to find a portrait of you.

I am glad to see what you look like. It is an exceedingly pleasant prospect! I have read your article "River Gold" with deep interest of course.

There are one or two points not quite accurate and as you will undoubtedly write and write about that country I must call your attention to them.

The first is about "Emory" Kolb. The name, by the way, is spelled Emery. It was Ellsworth, Emery's brother who navigated from <u>Wyoming to the sea</u>, but not continuously. And he was not the first. The first was Jack Sumner in 1869 with Andrew Hall. In 1890 R. B. Stanton did it too from Green River, Utah—and there are now one or two others I don't remember. Flavell in 1896 went from Green River, <u>Wyoming</u>, to Yuma. Nat Galloway went from Green River, <u>Wyo.</u>—to the Needles in 1896–97 winter.

Hite's gold was not the first taken out of the Colorado. The winter before we were to enter the Grand Canyon 1871–72 Major Powell was trying to find a place to take rations in, in the Kaibab section. At last he discovered that it could be done down Kanab Canyon. He went to the river that way with two or three men one of whom was George Riley, an Idaho prospector, who had taken a job with our party for the winter. He panned some gravel at the mouth of the Kanab and got a quantity of fine gold. This news got to Salt Lake and for several weeks there was a line of prospectors going to the Colorado via Kanab Canyon and out again. There was gold, but very fine "flour" gold which can be obtained anywhere along the Colorado. It is so fine it is hard to save it and working conditions are so difficult and space so limited that getting it out will never be profitable. There is always a chance of finding a pocket that is rich enough to pay but it is doubtful. The gold evidently comes mainly from the <u>shales</u> and rocks of Colorado and is so pulverized by erosion that it is like flour. This you state in your article.

I have some shirt studs, the tops made from gold Riley panned out on the Colorado—the result of amalgamating the fine gold gathered in a pan.

Your article is extremely interesting and valuable. I shall extract it from <u>Ax-I-Dent-Ax</u>—and file it with my Colorado River materials.

Thanks for the kind references to me.

Sincerely Yours

Frederick S. Dellenbaugh

May 26th, 1933

Dear Mr. Kelly:

Yours of the 22nd May is just here with the clipping about Capt. Bishop for which I thank you.

Bishop died on the 62nd anniversary of our start down Green River which was an interesting coincidence.

He was the last one of our party except me. He was not with us the second year as a nervous trouble developed because of the constant wetting and he was obliged to leave us in the winter of 1871–72.

We were warm friends all those 62 years. He was a fine character.

Certainly you can have a copy of the whole or any part of the translation of the Escalante Diary I had Mrs Bandelier make for me.

I will ask our manager here who is an expert typist how much he would charge and let you know.

It will be a fine thing for you and Birney to go over the trail, for that is the only way to get it right as to details. I will be glad to aid you all I can.

The reason Escalante went so far north was that he missed the trail at the La Sal Mts., so you will have to do the same.

Sincerely yours,

Frederick S. Dellenbaugh

July 23ᵈ 1934

Dear Mr. Kelly:

I suppose you wish I would write this on the machine but I must beg to use a pen which sometimes seems more intimate, easier and makes no noise.

You probably know by the papers if not otherwise that Dr. Frazier and his party are now bucking the rapids of the Grand Canyon. Clyde Eddy has joined them and as he has been as far as Bright Angell twice and once all the way through he ought to be a great help. I see by re-reading your letter you know it.

The river is lowest in many years and I think they will have a lot of hard work for more portages have to be made at low water. Eddy went through on high water and generally maintains that he is the only one who has done so, but we had extremely high water in 1872 and made as far as Kanab Canyon where, because of a combination of circumstances we were compelled to come out.

Mr. Winning is a young man from New York who is working this summer with a scientific group which is studying the Glen Canyon region, as far as possible, in detail. He thought he might have a chance to ride over the trail from Lee Ferry to the Ute Ford so I gave him all the data I had. As I did not hear from him in reply, I fear he left New York before receiving my letter.

I am positive that Escalante's crossing was not below the Ute Ford. It either was at the Ford or some distance above. His trail up to that point is perfectly clear. I surmise that if he missed the Ford it was by about a mile and a half and due to a severe storm that hit them just about where they should have turned to the right. They proceeded and passed down the small canyon 13 miles as the river runs, above the Ute Ford, but in a straight line much less than that. I used a league of 3 miles, based on our own three from the Paria mouth to the foot of the sand slope where the old trail went up, because they said it was one league over the same distance. If their league was shorter it would make some difference but not enough I think to change matters very much.

Perhaps, after all, you will be the one to determine the real crossing.

Minton Balch & Co. will send you a copy of your <u>Holy Murder</u> to autograph for me, and when you have done it please return to Cragsmoor, New York, as I shall be here, mainly, till about October 15th.

Yes Brown's Park (Hole) is a beautiful and interesting valley and the way the Green cuts into the mountains by the Gate of Lodore is superb. LODORE not LA-

DORE as some maps have it. I corrected the U.S.G.S. on that and future maps will be O.K. So also the post office. The canyon was named after Southey's poem, "How the waters come down at Lodore."

I don't understand how the U.S.G.S. made the mistake with the correct spelling in Powell's report and in my books. There was, at one time, a trading "fort" down near the Lodore entrance.

Yes Dr. Frazier has sent me a photograph of the tablet they are to place at Separation Rapid. It is a fine thing to do. While those men—in my opinion—are not entitled to a place on the "honor roll" of the Powell Monument at Grand Canyon, because anyway you look at it they were deserters, they ought to be commemorated in bronze too, and Separation Rapid is the place for it. Jack Sumner tried to dissuade them from leaving but, as he said, he was not so sure of his own side of the argument so he feared his plea was not very strong, furthermore, "the rapids had become a holy terror" to the elder Howland so there was no changing his mind.

"Shebit" is not correct. The Mormons pronounced it that way because unless one is very careful the "vw" sound in Pai Ute sounds like B. The correct spelling is SHEVWITS or SHIVWITS or Z. The reason Powell spelled it with an "i" instead of an "e" is that the Bureau of Ethnology devised an alphabet for spelling Indian words and in that an "i" is usually "e." I was among the Shevwits as you know and my spelling is as near their own pronunciation as can be. The Bureau spells tee-pee "tipi" and as a consequence many pronounce it "tippy." I still use "teepee" in ordinary writing. The "Hopi" should be Hopee. Many pronounce it Hoppy. Anyhow Hopee is one of the people and Hopeetuh all the people. They should be called Hopituh or Hopeetuh.

Sincerely yours

F. S. Dellenbaugh.

August 16th, 1934

Dear Mr. Kelly:

The book came through promptly and I must thank you most heartily for the delightful autographic compliment.

I have read here and there, being so deeply interested to get a sort of "bird's eye" view before taking it page by page as I shall now do.

You certainly "lambast" the Mormons "to beat the band" and there is much on your side, but there is also much on the other side. I have known the Mormons now for 63 years and many of them have been and are, close friends, so I look at them and their religion more sympathetically than you do. For my part I have never in my travels met a more kindly, helpful people. So there you are, Porter Rockwell notwithstanding.

The murder of Joseph Smith and his brother was a dastardly murder. I think you should have been more condemnatory of that. Of course, like the crucifixion of Jesus Christ, it provided a martyr, without which, no religion can flourish; but from the judicial standpoint it was a mean, cowardly, assassination as bad, or worse, than anything Porter Rockwell did. I have despised the crowd that did it ever since I first read about it.

Much that Brigham Young shouted from his pulpit was mere balderdash, couched in the language of God in that terrible book, the Old Testament, for the Bible was basic with the Mormons and unless one refused to believe it a divinely inspired book, he was immediately lost in any discussion of polygamy, murder or any other diabolical thing that those primitive Jews delighted to set down in their records.

Brigham Young's tirades were infantile compared with those of the Jews Divine leader. God says, Leviticus, XXVI–16– "I also will do this unto you: I will even appoint over you terror, consumption, and the burning ague, that shall consume the eyes, and cause sorrow of heart: and you shall sow your seed in vain, for your enemies shall eat it." "17. And I will set my face against you, and ye shall be slain before your enemies: they that hate you shall reign over you: and ye shall flee when none pursueth you" "22. I will also send wild beasts among you, which shall rob you of your children" and so on without limit of the fearful things, "I am the Lord, your God" (gentle creature) will do to anyone who opposes Him.

This kind of tirade was the same fanatical frenzy that Brigham invoked to keep his people in line, only he directed it more against his opponents. I am not trying to excuse the language of Brigham Young, only trying to show that it was more or less Biblical, and was not always real—only an effort to frighten people one way or the other, just as Jehovah's was—for of course it was the leaders of the Jews who spoke the words I have quoted not God at all. God is not like that.

So I feel more tolerance for the rantings of the Mormon leader than you do. Besides I doubt if Brigham Young sanctioned Porter Rockwell for I think he was too shrewd to do that. However I will probably know more about the subject than I do now when I have carefully read your interesting and valuable work—very important too in the literature concerning the religion of the Church of Jesus Christ of Latter Day Saints. I have read Linn's book—long ago. I have a copy and will read it again. But the fact remains that I like the Mormons. I have always found them kindly, friendly and helpful and I do not know a single instance where they were not helpful to Gentiles in distress barring Mt. Meadows of course as in the case of Fremont when he and his men straggled into Parowan used up, cold, hungry and on their last legs. The people of Parowan took them into their houses, fed them and sent them on their way rejoicing—rejoicing that there were in the world such kind and sympathetic people.

Jacob Hamblin told me that if he had been at home the Mountain Meadows Massacre would not have occurred and I believed him, for it was easy to see what a false move it was, leaving out the humanitarian side. Brigham Young would not have entertained such an act for a moment on purely political grounds if on no other.

Probably they were not particularly anxious to apprehend Lee because he had done good work for the Church over many years. Lee was a fanatic and so were Haight and Klingensmith, but Lee was not the worst of the trio—although his fanaticism ran higher on that terrible occasion. The murder of the people of that caravan was a frightful thing, but Haight, Lee & K had the murder of Joseph Smith and his brother as an example. That affair was still fresh in mind.

Many of the people who went out from Cedar on the occasion of the Mt. Meadows had no idea that murder was in the air. George Adair, whom I knew well, a young fellow at the time, said he joined the crowd without knowing what it was all about. Lee and the other two were the real perpetrators and I am sure they had no sanction from those higher up.

Lee protested to me that he really did not injure anyone and tried to prevent the slaughter but of course that was not true. He said he went home and cried like a baby—or that he cried and went home, and the Indians everafter called him Yahgats—cry baby.

There were bad men among the Mormons, of course, but the proportion was smaller than among the Gentiles.

There was one named Liston who lived in St. George in 1874—5 who was regarded by all the Mormons as a shifty character. Jacob told me that once, on the old trail down the Virgin when the Church had set up a post for the protection of the emigrants, a young man arrived travelling alone. Jacob came shortly after. Liston said "This man has got to go up"—meaning die. Jacob said, "No—I'll die first. Liston did not molest the man.

Well, I have written a lot. Now I'll read about Porter Rockwell. I have always thought he was a fine liar.

Sincerely yours

Frederick S. Dellenbaugh

P.S.

Off the Rockwell-Mormon subject on to Grand Canyon—beautiful Grand Canyon—dangerous Grand Canyon.

I hear that Frazier & party got through with only the loss of one boat. He is going to write me about it.

I have watched the papers here for news but there has been nothing in detail.
F.S.D.

It seems now that I shall not be able to get to Salt Lake or the West this year— yet the tide may turn yet.

I'd give a lot to see you in your own home—and some of my "Danite" Mormon friends whom I like and admire.
F.S.D.

[Much of the content of this letter is prompted by the book *Holy Murder, the Story of Porter Rockwell* (New York, 1934) written by Charles Kelly and Hoffman Birney. On the matter of the Mountain Meadows Massacre, one should consult the scholarly book by Juanita Brooks, *The Mountain Meadows Massacre* (Stanford, 1950, and Norman, Oklahoma, 1962).]

PART 5

The Lost Journal of John Colton Sumner

John Colton Sumner

May 24, 1869, ten men embarked at Green River, Wyoming, projecting a cruise through the canyons of the Green and Colorado Rivers. William H. Ashley, Denis Julien, William L. Manly, and H. M. Hook had boated sections of the Green River to total the course from the point of the 1869 embarkation to the confluence with the Colorado. Julien must be credited with thirty miles of the fast water of the Colorado immediately below the mouth of the Green. Except for the careful reporting by Ashley, information helpful to later navigators was fragmentary. The 1869 crew was led by Professor John Wesley Powell who requested his head boatman, John Colton Sumner, to write a journal. This document has had a confused history. A copy of that part applying to the cruise below the Uinta River is printed in volume 15 of the *Utah Historical Quarterly*. The editing is by William Culp Darrah who explained "When the three men deserted the party on August 28, 1869, they took with them one set of notes. The journals of Major Powell and Jack Sumner had been kept in duplicate, in case of loss or disaster, so that at least one record might be preserved for future use. In the haste and excitement of the hour the notes were not equally divided—rather both sets of the notes taken prior to July 2 were carried out by the elder Howland, and when the men were killed the notes were irretrievably lost." Fortunately the Sumner journal had been communicated to the world prior to the Howland tragedy.

On June 28, 1869, the party arrived at the mouth of the Uinta River and Professor Powell, William Rhodes Hawkins, and Frank Valentine Goodman walked to the Uintah Indian Agency on July 2. Goodman had lost his outfit, including his outer clothing, in the wreck of the *No Name* and was unable to replace it so remained at the agency. July 5 the party reassembled at the river and embarked the next day. Letters were sent to the agency, carried by some Indians, including one from Powell.

Mouth of Uinta River,
July 6, 1869.

Editors Missouri Democrat:

I send manuscript journal of one of the trappers con-
nected with the Colorado River Exploring Expedition.
I think you will find them somewhat lively, and may be
able to use them. Of course they will need "fixing" a little, may be toning somewhat. Jack Sumner, the writer,
has seen much wild life and read extensively. He has prepared the manuscript at my request. Should you conclude to publish he will send more.

Yours, &c,
J.W. POWELL
In charge Colorado Ex Ex.

Prefaced by Powell's letter, the journal was published in
the August 24 and 25, 1869, issues of the *Missouri Democrat*
under the caption "FROM COL. POWELL. Interesting memo-
randa. Daily Record of the Expedition. Latest Dates Published.
Journal of Jack Summers [*sic*] a Free Trapper in the Expedition, up to June 28th Inclusive." The following is the portion of
the Sumner journal published in the *Missouri Democrat*.

A Daily Journal of the Colorado Exploring Expedition

(Jack Sumner's Diary of the First Powell Expedition from Green River, Wyoming, to the Uinta Basin)

After many weeks of weary waiting, today sees us all ready for the adventures of an unknown country. Heretofore all attempts in exploring the Colorado of the West, throughout its entire course, have been miserable failures. Whether our attempt will turn out the same time alone can show. If we fail it will not be for the want of a complete outfit of material and men used to hardships. After much blowing off of gas and the fumes of bad whiskey, we were all ready by two o'clock and pulled out into the swift stream. The *Emma Dean*, a light four-oared shell, lightly loaded, carrying as crew Professor J. W. Powell, W. H. Dunn, and a trapper,[1] designed as a scouting party, taking the lead. The *Maid of the Canon* followed close in her wake, manned by Walter H. Powell and George Y. Bradley, carrying two thousand pounds of freight. Next on the way was *Kitty Clyde's Sister*, manned by as jolly a brace of boys as ever swung a whip over a lazy ox, W. H. Rhodes, of Missouri, and Andrew Hall, of Fort Laramie, carrying the same amount of freight. The last to leave the miserable adobe village was the *No Name* (piratic craft) manned by O. G. Howland, Seneca Howland, and Frank Goodman. We make a pretty show as we float down the swift, glossy river. As Kitty's crew have been using the whip more of late years than the oars, she ran on a sand-bar in the middle of the river, got off of that, and ran ashore on the east side, near the mouth of Bitter creek, but finally got off and came down to the rest of the fleet in gallant style, her crew swearing she would not "gee" or "haw" a "cuss." We moved down about seven miles and camped for the night on the eastern shore where there is a large quantity of cord wood. As it was a cold, raw night, we stole a lot of it to cook with. Proff., Walter, and Bradley spent a couple of hours geologising on the east side. Howland and Dunn went hunting down the river; returned at dark with a small sized rabbit. Rather slim rations for ten hungry men. The balance of the party stopped in camp, and exchanged tough stories at a fearful rate. We turned in early, as most of the men had been up for several proceding nights, taking leave of their many friends, "a la Muscovite." The natural consequence were fog[g]y ideas and snarly hair.

How strange it is that adopting foreign ways will so change us in many respects. If there is any meanness in a man, get him drunk and you soon see the Devil's claws, if not the whole of the traditional "Auld Cootie." If he is a goodhearted man when sober, he will be willing to sell his only shirt to help his friend. When I see how drink shows the true colors so plainly, I sometimes wish the whole world could be drunk for a short time, that the scoundrels might be all killed off through their own meanness.

May 25

Pulled out early, and dropped down to an old cabin, where I stole two bread-pans for the cook's use. Moved about eight miles and camped in the willows, as it was raining hard; stopped two hours; made some coffee, and cooked some villainous bacon to warm us up a little. Then pulled out again, as it showed some signs of clearing off. Went another five or six mile stretch when we saw five mountain sheep on a cliff; stopped to give chase, but they proved to be too nimble for us. Rhodes, however, found a lamb asleep on the cliff caught it by the heels and threw it off toward camp. The Professor and Bradley climbed a black looking cliff on the west side to see how it was made. All into camp by 3 o'clock, when we had our young sheep for dinner. Packed up the cooking utensils and pulled out again, and moved down through a rather monotonous country for six or eight miles further. Saw several wild geese and four beavers, but failed to get any. While rounding to on the west side, all the boats except Kitty's Sister got fast on a sand bar—the Maid so fast she had to be pried off with oars.[2] Camped on the west side, in the willow brush. While we were gathering drift-wood for camp fires, two mountain sheep ran out of the willows and up the side of the bluff. Two of the boys followed, but failed to get either of them. Rained all day and most of the night.

All afloat early; went about three miles, when we came to our first rapid. It cannot be navigated by any boat with safety, in the main channel, but the river being pretty high, it made a narrow channel, under the overhanging willows on the west shore, so that we were not delayed more than twenty minutes, all the boats but Kitty's Sister getting through easily. She getting on a rock, compelled Rhodes to get overboard and pry her off. About 4 o'clock, came to a meadow of about a thousand acres, lying between Green River and Henry's Fork. Camped for the night on the east shore, about a mile above the mouth of Henry's Fork. Passed the mouth of Black's Fork of the Green River today; it is but little wider at the mouth than at Fort Bridger, but deep. Henry's Fork is a stream about thirty feet wide, and is fed by the snows of the Uinta Mountains, about seventy-five miles northwest of this camp; it has some good pasturage on it, but no farming land, as it is at too great an altitude. At the mouth is a good place for one or two ranches. There are about three hundred acres of good land, but is inundated nearly every spring by freshets. There is a large stack of hay standing in the meadow, that has been left over from last year's crop.

Raised a cache that we made two months since, and found everything safe; moved down to the head of a canyon and camped on the east side, under a grove of cottonwood trees. Proff., Walter, and Bradley went geologising. Tramped around most of the day in the mud and rain to get a few fossils. Distance from Green River City to mouth of Henry's Fork sixty miles,[3] general course, 30 degrees E. of S; estimated land distance forty miles. Country worthless. Grease wood and alkali on the river bottom; on the hills sparse bunch grass, Artemissia, and a few stunted cedars. At intervals of four or five miles on the river there are a few scrubby cotton-woods, but none large enough for anything but fuel. Rained most of the day.

Still in camp. Proff. and the "Trapper" repaired a broken barometer. Walter and Bradley went geologizing on the west side. Bradley did not get into camp until night, having lost his way, and had a long, weary tramp through the mud and rain.

Proff climbed the hill on the east side of the canyon and measured it with a barometer; h[e]ight above the river 1140 feet, not perpendicular. There is a cliff on the west side that is fifty feet higher, and perpendicular. The rock is hard, fiery-red sandstone. It has been named Flaming Gorge.

Professor, Bradley, Senica, and Hall went up the river five miles, measuring a geological section. All in camp by three o'clock, when we loaded up and pulled on again into a channel as crooked as a street in Boston. Passed out of Flaming Gorge into Horseshoe Canon, out of Horseshoe Canon into Kingfisher Canon. While rounding a bend, we came on a herd of mountain sheep, that scampered up a steep, rocky side of the canon at an astonishing rate. The crews of the freight boats opened a volley on them that made the wilderness ring, reminding us all of other scenes and times, when we were the scampering party. Passed the mouth of a small stream coming in from the west, which we named King Fisher Creek, as there was a bird of that species perched on the branch of a dead willow, watching the finny tribe with the determination of purpose that we often see exhibited by politicians while watching for the spoils of office.[4] Killed two geese, and saw a great number of beavers today, but failed to get any of them. No sooner would we get within gun-shot, than down they would go with a plumping noise like dropping a heavy stone into the water. Made seven miles today, and camped for the night on the west bank opposite a huge grayish white sandstone that loomed up a thousand feet from the water's edge, very much the shape of an old-fashioned straw beehive, and we named it "Beehive Point." Saw the tracks of elk, deer and sheep on the sand. Near our camp, Goodman saw one elk, but missed it.

May 31

This morning Professor, Bradley, and Dunn went up the river two miles to examine some rocks and look for a lost blank book. Howland and Goodman climbed a high mountain on the west side to get a good view of the country at large, and so draw a good map. All ready by ten o'clock when we pull out and are off like the wind; ran about two miles through a rapid and into still water for half an hour, then to a bad rapid through which no boat can run; full of sunken rocks, and having a fall of about ten feet in two hundred yards. We were compelled to let our boats down along the west side with ropes from men holding the line, two men with oars keeping them off the rocks; made the passage in about two hours, and ran a large number of them in ten miles travel.

About 5 o'clock, we came to the worst place we had seen yet; a narrow gorge full of sunken rock, for 300 yards, through which the water run with a speed that threatened to smash everything to pieces that would get into it. All the boats were landed as quick as possible on the east side of the river, when we got out to examine the best point to get through, found ourselves on the wrong side of the river, and how to cross was the next question. We all plainly saw that it would be no child's play. Dunn and the trapper finally decided to take the small boat across or smash her to pieces; made the passage safe, unloaded and returned to relieve the freight boats, they taking out half their loads by making two trips with the freight boats and five with the small; we got everything safely across where we wanted it by sunset. Had supper; turned in, and in two minutes all were in dreamland.

June 1

After an early breakfast, all hands went to work letting the boats down with ropes, made the passage in three hours, when we jumped aboard again, and off we go like a shot; ran through about a dozen rapids in the course of ten miles, when we came to some signs of the country opening out. The walls were getting lower, and not so rough, and the current gradually slackens till it almost ceases. As the roaring of the rapids dies away above us, a new cause of alarm breaks in upon us from below. We ran along on the still water, with a vague feeling of trouble ahead, for about two miles, when, turning an abrupt corner, we came in sight of the first fall, about three hundred yards below us. Signaled the freight boats to land, when the *Emma* was run down within a rod of the fall, and landed on the east side. Her crew then got out to reconnoitre; found a fall of about ten feet in twenty-five. There is a nearly square rock in the middle of the stream about twenty-five by thirty feet, the top fifteen above the water. There are many smaller ones all the way across, placed in such a manner that the fall is broken into steps, two on the east side, three on the west. We all saw that a portage would have to be made here. Without any loss of time the *Emma Dean* was unloaded and pushed into the stream, four men holding the line, the remainder of the party stationed on the rocks, each with oar, to keep her from being driven on some sharp corners and smashed to pieces. Got her under the fall in fifteen minutes, when we returned, unloaded Kitty's Sister, had supper and went to sleep on the sand. There is not much of a canyon at the falls. Three hundred yards from the east side there is a cliff about 450 feet high, from whence the rocks have fallen to make the dam.

June 2

All out early to breakfast; dispatched it, and let Kitty's Sister over the falls as we did the small boat. Then came the real hard work, carrying the freight a hundred yards or more over a mass of loose rocks, tumbled together like the ruins of some old fortress. Not a very good road to pack seven thousand pounds of freight. Got the loads of the two boats over, loaded them, and moved down three hundred yards to still water; tied up and returned to the other boats, to serve them the same; got everything around in still water by 11 o'clock; had dinner and smoked all round; distance from Bee-hive Point unknown; course east of south; continuous canyon of red sand-stone; estimated height of one thousand feet; three highest perpendicular walls estimated at two thousand two hundred feet; named Red Canyon; on a rock [on] the east side there is the name and date—"Ashley, 1825"—scratched on evidently by some trapper's knife; all aboard, and off we go down the river; beautiful river, that increases its speed as we leave the fall, till it gets a perfect rapid all the way, but clear of sunken rocks; so we run through the waves at express speed; made seventeen miles through Red Stone canyon in less than an hour running time, the boats bounding through the waves like a school of porpoise. The *Emma* being very light is tossed about in a way that threatens to shake her to pieces, and is nearly as hard to ride as a Mexican pony. We plunge along singing, yelling, like drunken sailors, all feeling that such rides do not come every day. It was like sparking a black-eyed girl—just dangerous enough to be exciting. About three o'clock we came suddenly out to a beautiful valley about two by five miles in extent. Camped about the middle of it, on the west side, under two large pine trees; spread our bedding out to dry, while we rested in the shade. Two of the party came in at sunset, empty handed except the Professor, he being fortunate enough to get a brace of grouse. Spread our blankets on the clean, green grass, with no roof but the old pines above us, through which we could see the sentinel stars shining from the deep blue pure sky, like happy spirits looking out through the blue eyes of a pure hearted woman.

As we are guided on this voyage by the star in the blue; so may it be on the next, by the *spirit* in the blue.

June 3

Laid over to-day to dry out, and take observations. Several of the party hunting, but killed nothing. In the evening, some of the boys got out the fishing tackle and soon had the bank covered with queer mongrel of mackerel, sucker and whitefish; the other an afflicted cross of white fish and lake trout. Take a piece of raw pork and paper of pins, and make a sandwich, and you have the mongrels. Take out the pork and you have a fair sample of the edible qualities of the other kinds. From this camp to Bee-Hive Point is called by the Professor, Red Canyon, not very appropriately, as there are two distinct and separate canyons. This park is the best land we have seen, so far; good land; season long enough to raise rye, barley and potatoes, and all kinds of vegetables that would mature in four months. Irrigation not necessary, but if it should be, there is a beautiful clear trout stream running through the middle of it that can be thrown on almost any part of it at comparatively little cost.[5] Counting agriculture out, there is money for whoever goes in there and settles and raises stock. It is known by the frontiersmen as "Little Brown's Hole." Altitude is 6,000 feet. Game in abundance in the mountains south of the park; good trail to Green River City, and there could be a good wagon road made without a great outlay of money. All turn in early, as we want an early start in the morning.

June 4

All afloat early, feeling ready for anything after our rest. Had another splendid ride, of six or eight miles and came to the mouth of Red Fork, a most disgusting looking stream, coming in from the east, off of the "Bitter Creek Desert." It is about ten feet wide, red as blood, smells horrible and tastes worse. Passed on through five miles more of canyon and came to "Brown's Hole," a large valley, about twenty miles long and five wide—splendid grass on it.[6] Passed on about the middle of the valley and camped at the mouth of a small trout stream, coming in from the east, named on Fremont's map, "Tom Big Creek." Had dinner and moved down about two miles, and camped on the west side of Green river, under a great cotton-wood tree that would furnish shade and shelter for a camp of two hundred men. Hall killed several ducks in a lake near camp, and in the evening Bradley, Howland and Hall caught a large number of fish.

June 5

This morning we were all awakened by the wild birds singing in the old tree above our heads. The sweet songs of birds, the fragrant odor of wild roses, the low, sweet rippling of the ever murmuring river at sunrise in the wilderness, made everything as lovely as a poet's dream. I was just wandering into paradise; could see the dim shadow of the dark-eyed houris, when I was startled by the cry, "Roll out; bulls in the corral; chain up the gaps"—our usual call to breakfast. The hour is vanished, and I rolled out to fried fish and hot coffee. The Professor and Dunn climbed the hill south of camp, two miles from the river—h[e]ight, 2200 feet; Howland spent the day dressing up his maps ; Bradley, Seneca and Hall crossed to eastside and measured off a geological section. The remainder of the party spent the day as best suited them. Measured the old tree; circumference, 5 feet from the ground, 23 ½ feet.

June 6

Took our time in getting off, as we had but a short journey before us for the day; but it proved a pretty hard one before we got done with it. No sooner had we started than a strong head-wind sprung up directly in our faces. Rowed about twenty-five miles against it—no easy task, as the river is a hundred and fifty yards wide, with hardly any current. Saw thousands of ducks of various kinds; killed a few, and one goose. Camped at the head of a canyon, at the southern end of the valley, on the east side of the river under a grove of box elder trees. The Professor and Hall caught another mess of fish.

June 7

Still in camp. Professor and Dunn measured the wall of the canyon on the east side; Bradley geologising; Howland and Goodman sketching; Rhodes brushing up Kitty's Sister, swearing all the time that she can stand more thumps than Kitty ever could. Professor and Dunn came in at noon and reported the wall 2086 feet. All the party in camp the rest of the day; wind and rain in the evening. Distance from the mouth of Henry's Fork 90 miles; general course of the river 25 degrees south of east.[7] The valley called Brown's Hole is a pretty good piece of land; would make a splendid place to raise stock; it has been used for several years as a winter herding ground for the cattle trains. Last winter there were about 4000 head of oxen pastured in it without an ounce of hay. I saw them in March and am willing to swear that half of them were in good enough order for beef.

June 8

Pulled out early and entered into as hard a day's work as I ever wish to see. Went about a half a mile when we came to a terrible rapid, and had to let our boats down with ropes. Passed about a dozen bad rapids in the forenoon. Camped for dinner on the east side at the foot of a perpendicular rose-colored wall, about fifteen hundred feet; pulled out again at one o'clock; had proceeded about half a mile when the scouting boat came to a place where we could see nothing but spray and foam. She was pulled ashore on the east side and the freight boats instantly signaled to land with us. The Maid and Kitty's Sister did so but the No Name being too far out in the current and having shipped a quantity of water in the rapid above, could not be landed, though her crew did their best in trying to pull ashore at the head of the rapid, she struck a rock and swung into the waves sideways and instantly swamped. Her crew held to her while she drifted down with the speed of the wind; went perhaps 200 yards, when she struck another rock that stove her bow in; swung around again and drifted toward a small island in the middle of the river; here was a chance for her crew, though a very slim one. Goodman made a spring and disappeared; Howland followed next, and made the best leap I ever saw made by a two-legged animal, and landed in water where he could touch the rocks on the bottom; a few vigorous strokes carried him safe to the island. Seneca was the last rat to leave the sinking ship, and made the leap for life barely in time; had he stayed aboard another second we would have lost as good and true a man as can be found in any place. Our attention was now turned to Goodman, whose head we could see bobbing up and down in a way that might have provoked a hearty laugh had he been in a safe place. Howland got a pole that happened to be handy, reached one end to him and hauled him on the isle. Had they drifted thirty feet further down nothing could have saved them, as the river was turned into a perfect hell of waters that nothing could enter and live. The boat drifted into it and was instantly smashed to pieces. In half a second there was nothing but a dense foam, with a cloud of spray above it, to mark the spot. The small boat was then unloaded and let down with ropes opposite the wrecked men on the island. The trapper crossed over and brought them safely to shore to the east side. She was then let down about a half a mile further, where we could see part of the stern cuddy of the wrecked boat on a rocky shoal in the middle of the river. Two of the boys proposed to take the small boat over and see how much of the lost notes could be recovered. The Professor looked ruefully across the foaming river, but forbade the attempt. All hands returned to the head of the rapids, feeling glad enough that there were no lives lost, a little sore at the loss of the boat and cargo of 2,000 pounds of provisions and ammunition, all the personal outfit of the crew, three rifles, one revolver, all the maps and most of the notes and many of the instruments, that cannot be replaced in time to carry on the work this year, and do it right, ate supper of bread and bacon, and went to sleep under the scrubby cedars.

June 9

All up by sunrise and at work unloading the boats, ready for letting down with ropes. Got the boats and remaining cooking utensils over and opposite the wreck on the shore. Had dinner, when Hall bantered one of the men to go over to the wreck and see what there was left. Away they went and got to it safely, after a few thumps on the rocks, and fished out three barometers, two thermometers, some spare barometer tubes, a pair of old boots, some sole leather, and a ten gallon cask of whisky that had never been tapped. Not a sign of anything else. How to get back was the next question, it being impossible to go back over the route they came. A narrow, rocky race offered a chance to get through the island into the main channel. After an hour's floundering in the water among the rocks, they got through to the main channel, and dashing through some pretty rough passes, they reached the shore, where the rest of the party stood ready to catch the lines, their arms extended, like children reaching for their mother's apron strings. The Professor was so much pleased about the recovery of the barometers, that he looked as happy as a young girl with her first beau; tried to say something to raise a laugh, but couldn't. After taking a good drink of whisky all around, we concluded to spend the rest of the day as best suited. Some packed freight for future use; the rest slept under the shade of the scrubby cedars.[8]

June 10

Out early again, and at work carrying the rest of our freight over the land. Had all done by noon; eat dinner, loaded up, and let down another two hundred yards with ropes, when we got aboard and rowed about half a mile. Crossed over to west side and let down another rapid through a narrow race. *Emma* and the Maid passed through safe but poor Kitty's Sister got a hole stove in her side. Camped for the night on the west side, on the sand.

June 11

Rapids and portages all day. By hard work we made three miles. Passed the mouth of a small trout stream coming in from the west. Hall shot an osprey on her nest in the top of a dead pine, near the mouth of the creek. Camped for the night on the west side, under an overhanging cliff.

June 12

More rapids that are impossible to run. Excessively hard work. Made three miles. Camped on the east side of the river in a grove of elder trees, at the head of a long rapid. Will have to make another half mile cartage.

June 13

Rested today, as we all need it very much. Three of the boys went hunting, but there is nothing in this part of the country but a few mountain sheep, and they stay where a squirrel could hardly climb. We are looking for better traveling pretty soon, as we have got to the point where the white sandstone caps the hard red, 2000 feet above our heads.

June 14

Still in camp; repaired a broken barometer and started our plunder over the Portage; Professor and Howland have been busy for two days restoring the lost maps.

June 15

Made the Portage and ran a bad rapid at the lower end of the Portage and through half a mile of smooth water, when we came to another impassable rapid; unloaded boats and camped on the east side, under some scrubby cedars. Rain at dark. Made a trail and turned in for the night.

June 16

Pulled out early and went to work with a will. While letting the Maid down with ropes she got crossways with the waves and broke loose from the five men holding the line, and was off like a frightened horse. In drifting down she struck a rock that knocked her stern part to pieces. Rhodes and the trapper jumped in the small boat and gave chase; caught her half a mile below. Got everything over by sunset and camped on the east side on a sand bar. Pulled out at seven o'clock, and ran a bad rapid the first half mile. The freight boats went through in good style, but the *Emma*, in running too near the east shore, got into a bad place and had a close collision, filling half full, but finally got out, all safe, baled out, and ran two miles through smooth water, when we came to another bad rapid and had to let down with ropes; tied up while we repaired the Maid; passed the rapid and on through two miles of smooth water, when we came to another rapid that has a fall of about twelve feet in a hundred and fifty, but clear of rock; the *Emma* ran through without shipping a drop, followed close by the Maid, she making the passage without shipping much, but poor Kitty's Sister ran on a rock near the east side and loosened her head block and came down to the other boats leaking badly. She was run ashore when Rhodes caulked her with some oakum that would serve to keep her afloat for a while; when we pulled out again, run half a mile with the *Emma* and got into a complete nest of whirlpools, and got out of them by extreme hard work; decided that it was unsafe for the freight boats to attempt it; so we were compelled to let them down with ropes on the east side through a narrow channel. Jumped aboard again and pulled down two miles further through smooth water, and camped for the night on the east side at the head of a rapid. While we were cooking supper a whirlwind came up the canyon, and in an instant the fire was running everywhere; threw what happened to be out on board again as quickly as possible, and pulled out and ran the rapid and on down two miles further and camped again on the east side, and commenced anew our preparations for supper; had supper, and laughed for an hour over the ludicrous scene at the fire. Went to bed and were lulled to sleep by the rain pattering on the tent.

June 18

Repaired Kitty's Sister and pulled out again, and had a splendid ride of six miles, and came to the mouth of Bear River,[9] a stream one hundred and twenty yards wide and ten feet deep; camped on a point of land between the two rivers, under some box elder trees. All hands went to work fishing, and soon had a good number of them. Bradley was much provoked by one large one that carried off three of his best hooks, but finally got him with a strong line got up for his especial benefit. He proved to be about thirty inches long and fifteen pounds weight. Opposite the mouth of Bear River there is the prettiest wall I have ever seen. It is about three miles long and five hundred feet high, composed of white sandstone, perpendicular and smooth, as if built by man. It has been christened Echo Rock, as it sends back the slightest and most varying sounds that we can produce.[10]

June 19

Still in camp. Professor, Bradley and Hall climbed the northern end of Echo Wall. The remainder of the men in camp, fishing, washing, etc.

June 20

All hands in today, taking a general rest. Wrote our names on Echo rocks opposite the camp. The entire distance from the southern end of the valley called Brown's Hole to the mouth of the Bear river is a canyon, except at two creeks on the west side, where there is a gorge cut through by the water of each. It has been named Ladore canyon by the Professor, but the idea of diving into musty trash to find names for new discoveries on a new continent is un-American, to say the least. Distance through it, 25 miles; general course, 25 degrees west of south; average h[e]ight on both sides about 1700 feet; highest cliff measured (Black Tail Cliff) 2307 feet. There are many still higher but having enough other work on our hands to keep us busy, we did not attempt to measure them.[11]

June 21

Off at seven o'clock and row down for one mile and a half along the base of Echo Wall, a nearly south course; passed the point of it, turned and ran due north for about five miles; back into the hard, red sandstone again, through a narrow, dangerous canyon full of whirlpools, through which it is very hard to keep a boat from being driven on the rock; if a boat should be wrecked in it her crew would have a rather slim chance to get out, as the walls are perpendicular on both sides and from 50 to 500 feet high. Made a portage at the lower end of it; had dinner and pulled out again, and went five miles further, making one short portage on the way; camped for the night on the west side, at the mouth of a clear, beautiful trout stream. Mr. Howland dropped his maps and pencils, rigged a line, and soon had a score of large trout, the first we have been able to catch so far. Made fifteen miles to-day; continuous canyon, named "White Pool Canyon"; trout stream named Brush Creek.[12]

June 22

After a good breakfast of fried trout, we pulled out and made a splendid run of six miles through a continuous rapid and stopped to have a hunt, as we saw many tracks of deer and sheep on the sand. All ready by one o'clock, when the *Emma* started down a long rapid, which has a fall of about thirty feet per mile. Went along in splendid style till she got to the lower end, where there is a place about a hundred yards long that had a dozen waves in it fully ten feet high. As she could not be pulled out of there her crew kept her straight on her course and let her ride it out. Went through them safe, but shipped nearly full, and pulled ashore looking like drowned rats. Decided it unsafe for the freight boats to try it, so we were compelled to make a short portage and let down with ropes. Jumped aboard again and pulled out into more rapids, every one of which would thoroughly drench us and leave an extra barrel or two in the boats; but we kept bailing out without any unnecessary stoppages. Dancing over the waves that had never before been disturbed by any keel, the walls getting gradually lower, till about four o'clock, when we came suddenly out into a splendid park; the river widened out into a stream as large as the Missouri, with a number of islands in it covered with cottonwood trees. Camped on the first one we came to, and rolled out on the grass in the shade to rest. Distance from mouth of Bear River 26 miles.[13] General course 20 degrees south of west from Brown's Hole to this point. The whole country is utterly worthless to anybody for any purpose whatever, unless it should be the artist in search of wildly grand scenery, or the geologist, as there is a great open book for him all the way.

June 23

Unloaded the boats and spread our plunder out to dry. Rhodes, Dunn, Harding [?], Goodman, and Howland sketching; the others, in common, repairing boats and washing. Hunter came in about noon with a fine buck that Rhodes had killed, when we loaded up and moved down about five miles, and camped on the east side, at the lower end of a splendid island covered with a heavy growth of cottonwood. Our camp is within half a mile of the last one above, the river making an almost complete circle.

June 24

The Professor and Howland climbed the mountain on the east side, with barometer and drawing materials; spent the greater part of the day sketching the park, which they have named "Island Park"; rain in the evening.

June 25

Pulled out at seven and moved down four miles, to the head of another canyon—cragged canyon—and into more rapids; made two portages and camped on the west side, at the head of another impassable rapid to loaded boats; one of the men sick.[14]

June 26

Made the portage and went a short distance when we came to another one, and had to make it in the rain; while the men were at work the Professor climbed up the side-hills looking for fossils; spent two hours to find one, and came back to find a peck that the men had picked up on the bank of the river; all ready by three o'clock, when we pulled out again; ran four miles at a rapid rate through the canyon, when all at once the Great Uinta Valley spread out before us as far as the eye could reach. It was a welcome sight to us after two weeks of the hardest kind of work, in a canyon where we could not see half a mile, very often, in any direction except straight up. All hands pulled with a will, except the Professor and Mr. Howland. The Professor being a one-armed man, he was set to watching the geese, while Howland was perched on a sack of flour in the middle of one of the large boats, mapping the river as we rowed along. Our sentinel soon signaled a flock of geese ahead, when we gave chase, and soon had ten of them in the boats. Summed up the log, found we had run 23 miles since leaving the canyon, and camped for the night on the east side, under three large cottonwoods. Rested, eat supper, and turned in to be serenaded by the wolves, which kept up their howling until we dropped asleep, and I don't know how much longer, as I heard them next morning at daybreak.

June 27

Off again at seven, down a river that cannot be surpassed for wild beauty of scenery, sweeping in great curves through magnificent groves of cotton wood. It has an average width of two hundred yards and depth enough to float a New Orleans packet. Our easy stroke of eight miles an hour conveys us just fast enough to enjoy the scenery, as the view changes with kaleidescopic rapidity. Made sixty three miles today, and camped on the west side, at the mouth of a small, dirty creek. Killed eight wild geese on the way.

June 28

Same character of country as yesterday. Saw four antelope, but failed to get any. Forty-eight miles brought us to the mouth of Uinta river, which place we reached about three o'clock, and camped on the west side of Green river, under a large cottonwood, at the crossing of the Denver City and Salt Lake wagon road, as it was located in 1865. There is not much of a road now, if any, as it has never been traveled since unless by wolves, antelope, and perhaps a straggling Indian at long intervals. Distance from the head of the valley by river, one hundred and thirty miles; by land about fifty miles.[15] General course of the river 10° west of south. This part of the country has been written up so often by abler pens, that I hesitate in adding anything more. As an agricultural valley it does not amount to much, as it is too dry on the uplands, and there are but few meadows on the river bottom, and they as a general rule are small—from fifty to two hundred acres in extent: The only exception that I know of is one opposite our present camp, lying between Green and White rivers. It is about two thousand five hundred acres in size, and overflows, though very seldom. At present it is clothed with a thick growth of grass, waist high. On the uplands there is the common bunch grass of the west—short but very rich. No part of the country that we have seen can be irrigated, except the river bottoms, as the uplands are rolling and cut up by ditches in almost every direction. But for a stock country it would be hard to excel, as almost all kinds would do well on the bunch grass throughout the entire year. There is plenty of timber for building purposes and fuel, and enough farming land to produce all that a large settlement would require for home consumption. But there is one thing in the way. According to the treaty of 1868 between Gov. Hunt, of Colorado, and the Ute Indians, most, if not the whole of this valley belongs to the reservation, selected by the Indians themselves. Whether they will be permitted to keep it or not remains to be seen. Most likely they will, as one band of them have a permanently settled thing of it, and have a winter agency twenty-five miles from this point on Uinta river. What the country is below I know not. As far as the eye can reach there is a rolling prairie with a dark line through it that marks the course of the Green River. It is reasonable to suppose it to be the same character of country as that we have passed through in our last two days' travel. So far we have accomplished what we set out for. We were told by the frontiersmen while at Green River that we could not get to the mouth of White River. One man that filled the important office of police-man in Pgitmont had the assurance to tell me that no boat could get as far down as Brown's Hole. We expect to remain here for a week to meet Col. Mead, and send off some specimens and all the notes and maps, to make sure of that much. Total distance run 356 miles; estimated distance to junction and Grand rivers 300 miles by river.[16]

List of Animals Living in the Country through Which We Have Passed

Grizzly bear, cinamon bear, black bear, elk, mule deer, mountain sheep or big-horn, prong-horned antelope, gray wolf, prairie wolf, cougar, red fox, marten, mink, lynx, wild cat, prairie dog, beaver, otter, muskrat, badger, ground hog, mountain rat, gray prairie squirrel, large striped ground squirrel, small do. do., small shrews and mice.

List of Birds Seen on the Way

Wild geese, ducks of almost every kind, loon, stork, bittern, cormorant, rails, woodcock, snipes of many kinds, curlew, osphey, pelican, sand hill crane, bald eagle, golden eagle, colored raven, common crow, Clark's crow, sage grouse, black grouse, short-tailed grouse, magpie, long-crested jay, Canada jay, light blue jay, red-shafted flicker, small blackbirds, red-winged starling, Southern mocking bird, robin, brown thrush, cross-beak, wren, sparrows, sparrow-hawk, sharped-shinned hawk, roose hawk, pigeon hawk, mourning dove, meadow lark, woodpeckers of all kinds, and buzzards.

I write this at the request of Professor Powell, he urging me from the begin-ning to do so, while I, knowing there were many able pens in the party, as persis-tently declined, till I could no longer do so with any show of reason. I have written this with many misgivings, being more used to the rifle, lariat and trap, than the pen. Receiving no hints from any one, I have been compelled to write as I could. Were I to study grammar a little and sacrifice truth to flights of fancy, I might make a more interesting report, but I shall let it stand as it is. If it meets the approval of the public, well and good; if it does not, I will leave the report of the rest of the trip to other and abler hands, and return to my rifle and trap.

Jack Sumner
Free Trapper

Notes

1. The title of trapper was of high standing from 1824 to 1840 when the mountain men were stripping the fur-bearing animals from the Upper Colorado River Basin, and Sumner was accepting this status in the third person.

2. The frequent groundings, the extensive linings, and the laborious portages attest to the poorly designed hulls and the inexperience of the crew.

3. The distances traveled were estimated, and Sumner's guess of sixty miles was close to the sixty-eight miles shown by the 1922 survey.

4. The stream later acquired the name Sheep Creek. Beehive Point is opposite.

5. The camp was in Little Hole which is cut by Little Davenport Creek.

6. Brown's Hole, whose inhabitants prefer the name Brown's Park, is named for its first resident, Baptiste Brown, who arrived there in 1827. Other sources give his arrival in 1835, and he made his headquarters there until 1843. LeRoy Hafen believes he is probably the same person known as Jean Baptiste Chalifou. B. Chalifou is inscribed with the date 1835 in the Willow Creek drainage below Ouray.

7. Sumner's guess of ninety miles from the mouth of Henrys Fork to the Gate of Lodore was sixteen miles too much.

8. In addition to the soaking in the river, there is much wallowing in fiction in this description of the No Name wreck but no word pictures equal in exaggeration the Moran drawing appearing as Figure 10 in J. W. Powell's *Exploration of the Colorado River of the West . . .* (Washington, D.C., 1875). The incompetent planning, improper equipment design, and the lack of necessary skills had produced the inevitable adventure.

9. The Bear River is now the Yampa.

10. Echo Rock is also known as The Blade and Steamboat Rock.

11. The Canyon of Lodore was measured as eighteen and one-half miles by the 1922 survey.

12. Obviously the journal read Whirlpool Canyon. Brush Creek is now Jones Hole Creek.

13. Bear River to Island Park is eleven miles.

14. Sumner was the sick man.

15. The river miles from the foot of Craggy Canyon, now known as Split Mountain, to the mouth of the Uinta are 71 ½.

16. They had traveled 258 ½ miles and had 245 miles more cruising to the mouth of the Green River.

Permissions and Credits

Index

Indians: Hillers's photographs of, 35, *144–52, 157–60*, 168; and Powell as Special Commissioner of Indian Affairs, 37. *See also* archaeology; Hopi; Indian Territory; language; Moqui; Navajo; Paiute; Snake; Southern Paiute; Shivwits; Uintah; Ute; White River Reservation

Indian Territory, 121, 168, 173n86. *See also* Cherokee Nation; Cheyenne; Creek; Seminole

Ingalls, George W., 37, 173n87

Island Park, 15, 53, 239

Ives, Joseph C., 193, 201

Jackson, William H., 35

Jarvis, J. F., 169n13

Johnson, William Derby, Jr., 87, 95, 172n50

Johnson Canyon, 95, 116

"John Wesley Powell's Journal: Colorado River Exploration, 1871–72" (*Smithsonian Journal of History*, vol. 3, Summer 1968), xiv

Jones, John Buttrick, 123, 174n95

Jones, Stephen Vandiver, xix, 4, 40, 178

Jones Hole Creek, 30n48, 170n15, 242n12

Jones Peak, 70

journals, of Powell expeditions: and documentation of 1871 expedition, 5–6; editing of for publication, 6, 40; and Hillers's diary of 1871 expedition, 39–40, 41–129; and letters of Dellenbaugh, 196–221; and Powell's account of 1871 expedition, 7–27; reprinting of by University of Utah Press and Utah State Historical Society, xiv; and Sumner on 1869 expedition, 225, 227–41

Julien, Denis, x, 195, 205, 207, 225

Jumper, John, 123, 174n94

Kaibab Paiute Indians, *144, 145, 146, 149*, 173n81. *See also* Paiute; Southern Paiute

Kaibab Plateau, 109

Kanab (settlement), 81

Kanab Canyon, *143*

Kanab Creek, xiii, 27

Kanab Gap, 85

Kanab River, 81

Kanab Wash, 114

Kane Canyon, 213

Kane Creek, 202. *See also* Cane Creek

Kannarah (Kannaraville), 83

Kelly, Charles, 194, 195, 205, 207, 221

Kettle Creek, 44

Kickapoo (Indian Territory), 125

King, Clarence, 38

Kingfisher Canyon, 8, 9, 28n20, 43, 230

Kingfisher Creek, 8, 9, 43, 230

kiva, 22, 31n74

Klingensmith, Philip, 220

Kolb, Ellsworth, 215

Kolb, Emery, 209, 215

Kolob Plateau, *139*

Kwagunt Creek, 31n83

Kwagunt Valley, 25

Labyrinth Alcove, 23, 31n76

Labyrinth Canyon, 17, 71, 72

Labyrinth Park, 17

Lake Creek, 98

Lake Mountain (Aquarius Plateau), 97

languages, Powell's lexicons of Native American, xiii. *See also* Tewa language

Las Vegas Paiute, *150, 151*

Lava Cliff, xi

Layton, Thomas, 57

Leaping Brook, 13, 50

Lee, Emma, 106, 172n65, 212

Lee, John D., 95, 106, 107, 117, 171n30, 172n72, 200, 207, 212–13, 220

Lee, Lavinia Young, 95

Lee, Rachel, 117, 172n65, 212

Lee's Ferry, 24

Lena, Mt., 10, 46

Len's Canyon, 31n66

Lewis and Clark expedition (1803–05), 36

Lignight Canyon, 68

literature, on Powell expeditions: and role of Dellenbaugh as chronicler, xiii; in scholarly journals and popular media, xiv. *See also* journals

Little Brown's Hole, 232, 242n5

Little Colorado River, 25, 109

Little White River, 68

Lodore, Gate of, 47–48, 217–18, 242n7. *See also* Canyon of Lodore

Log Cabin Cliff, 64

Long Valley, 27

Loper, Bert, 170n23

Loper Ruin, 31n74, 170n23

"The Lost Journal of John Colton Sumner" (*Utah Historical Quarterly* 1969), xiv

Lunt, Henry, 84

Manley, William L., 225

Manti (town), 82

maps and map-making: importance of to Powell expeditions, 177; of Indian Territory, *121*; reproduction of Bishop's for Green and Colorado Rivers, *xiii*, 178, 179–88. *See also* topography

Marble Canyon, 108, *142*

Marble Pinnacle, *143*